Neural Networks in Finance: Gaining Predictive Edge in the Market

Neural Networks in Finance: Gaining Predictive Edge in the Market

Paul D. McNelis

ELSEVIER
ACADEMIC
PRESS

Amsterdam • Boston • Heidelberg • London • New York • Oxford
Paris • San Diego • San Francisco • Singapore • Sydney • Tokyo

Elsevier Academic Press
30 Corporate Drive, Suite 400, Burlington, MA 01803, USA
525 B Street, Suite 1900, San Diego, California 92101-4495, USA
84 Theobald's Road, London WC1X 8RR, UK

This book is printed on acid-free paper. ∞

Library of Congress Cataloging-in-Publication Data
McNelis, Paul D.
  Neural networks in finance : gaining predictive edge in the market / Paul D. McNelis.
    p. cm.
  1. Finance–Decision making–Data processing. 2. Neural networks (Computer science) I. Title.
  HG4012.5.M38 2005
  332′.0285′632–dc22

                    2004022859

British Library Cataloguing in Publication Data
A catalogue record for this book is available from the British Library

ISBN: 0-12-485967-4

For all information on all Elsevier Academic Press publications
visit our Web site at www.books.elsevier.com

Printed in the United States of America
04   05   06   07   08   09   9   8   7   6   5   4   3   2   1

Contents

II Applications and Examples   **113**

Preface

Adjusting to the power of the Supermarkets and the Electronic Herd requires
a whole different mind-set for leaders ...

Thomas Friedman, *The Lexus and the Olive Tree*, p. 138

Questions of finance and market success or failure are first and foremost *quantitative*. Applied researchers and practitioners are interested not only in predicting the *direction of change* but also *how much* prices, rates of return, spreads, or likelihood of defaults will change in response to changes in economic conditions, policy uncertainty, or waves of bullish and bearish behavior in domestic or foreign markets. For this reason, the premium is on both the *precision* of the estimates of expected rates of return, spreads, and default rates, as well as the *computational ease* and *speed* with which these estimates may be obtained. Finance and market research is both empirical and computational.

Peter Bernstein (1998) reminds us in his best-selling book *Against the Gods,* that the driving force behind the development of probability theory was the *precise* calculation of odds in games of chance. Financial markets represent the foremost "games of chance" today, and there is no reason to doubt that the precise calculation of the odds and the risks in this global game is the driving force in quantitative financial analysis, decision making, and policy evaluation.

Besides precision, speed of computation is of paramount importance in quantitative financial analysis. Decision makers in business organizations or in financial institutions do not have long periods of time to wait before having to commit to buy or sell, set prices, or make investment decisions.

While the development of faster and faster computer hardware has helped to minimize this problem, the specific way of conceptualizing problems continues to play an important role in how quickly reliable results may be obtained. Speed relates both to computational hardware and software.

Forecasting, classification of risk, and dimensionality reduction or distillation of information from dispersed signals in the market, are three tools for effective portfolio management and broader decision making in volatile markets yielding "noisy" data. These are not simply academic exercises. We want to forecast more accurately to make better decisions, such as to buy or sell particular assets. We are interested in how to measure risk, such as classifying investment opportunities as high or low risk, not only to rebalance a portfolio from more risky to less risky assets, but also to price or compensate for risk more accurately.

Even in a policy context, decisions have to be made in the context of many disparate signals coming from volatile or evolving financial markets. As Othmar Issing of the European Central Bank noted, "disturbances have to be evaluated as they come about, according to their potential for propagation, for infecting expectations, for degenerating into price spirals" [Issing (2002), p. 21].

How can we efficiently distill information from these market signals for better diversification and effective hedging, or even better stabilization policy? All of these issues may be addressed very effectively with neural network methods. Neural networks help us to approximate or "engineer" data, which, in the words of Wolkenhauer, is both the "art of turning data into information" and "reasoning about data in the presence of uncertainty" [Wolkenhauer (2001), p. xii]. This book is about predictive accuracy with neural networks, encompassing forecasting, classification, and dimensionality reduction, and thus involves data engineering.[1]

The benchmark against which we compare neural network performance is the time-honored linear regression model. This model is the starting point of any econometric modeling course, and is the standard workhorse in econometric forecasting. While there are doubtless other nonlinear methods against which we can compare the performance of neural network methods, we choose the linear model simply because it is the most widely used and most familiar method of applied researchers for forecasting. The neural network is the nonlinear alternative.

Most of modern finance theory comes from microeconomic optimization and decision theory under uncertainty. Economics was originally called the "dismal science" in the wake of John Malthus's predictions about the relative rates of growth of population and food supply. But economics can be dismal in another sense. If we assume that our real-world observations

---

[1]Financial engineering more properly focuses on the design and arbitrage-free pricing of financial products such as derivatives, options, and swaps.

come from a linear data generating process, that most shocks are from an underlying normal distribution and represent small deviations around a steady state, then the standard tools of classical regression are perfectly appropriate. However, making use of the linear model with normally generated disturbances may lead to serious misspecification and mispricing of risk if the real world deviates significantly from these assumptions of linearity and normality. This is the dismal aspect of the benchmark linear approach widely used in empirical economics and finance.

Neural network methods, coming from the brain science of cognitive theory and neurophysiology, offer a powerful alternative to linear models for forecasting, classification, and risk assessment in finance and economics. We can learn once more that economics and finance need not remain "dismal sciences" after meeting brain science.

However, switching from linear models to nonlinear neural network alternatives (or any nonlinear alternative) entails a cost. As we discuss in succeeding chapters, for many nonlinear models there are no "closed form" solutions. There is the ever-present danger of finding locally optimal rather than globally optimal solutions for key problems. Fortunately, we now have at our disposal evolutionary computation, involving the use of genetic algorithms. Using evolutionary computation with neural network models greatly enhances the likelihood of finding globally optimal solutions, and thus predictive accuracy.

This book attempts to give a balanced critical review of these methods, accessible to students with a strong undergraduate exposure to statistics, econometrics, and intermediate economic theory courses based on calculus. It is intended for upper-level undergraduate students, beginning graduate students in economics or finance, and professionals working in business and financial research settings. The explanation attempts to be straightforward: what these methods are, how they work, and what they can deliver for forecasting and decision making in financial markets. The book is not intended for ordinary M.B.A. students, but tries to be a technical exposé of a state-of-the-art theme for those students and professionals wishing to upgrade their technical tools.

Of course, readers will have to stretch, as they would in any good challenging course in statistics or econometrics. Readers who feel a bit lost at the beginning should hold on. Often, the concepts become much clearer when the applications come into play and when they are implemented computationally. Readers may have to go back and do some further review of their statistics, econometrics, or even calculus to make sense of and see the usefulness of the material. This is not a bad thing. Often, these subjects are best learned when there are concrete goals in mind. Like learning a language, different parts of this book can be mastered on a need-to-know basis.

There are several excellent books on financial time series and financial econometrics, involving both linear and nonlinear estimation and forecasting methods, such as Campbell, Lo, and MacKinlay (1997); Frances and van Dijk (2000); and Tsay (2002). In additional to very careful and user-friendly expositions of time series econometrics, all of these books have introductory treatments of neural network estimation and forecasting. This work follows up these works with expanded treatment, and relates neural network methods to the concepts and examples raised by these authors.

The use of the neural network and the genetic algorithm is by its nature very computer intensive. The numerical illustrations in this book are based on the MATLAB programming code. These programs are available on the website at Georgetown University, www.georgetown.edu/mcnelis. For those who do not wish to use MATLAB but want to do computation, Excel add-in macros for the MATLAB programs are an option for further development. Making use of either the MATLAB programs or the Excel add-in programs will greatly facilitate intuition and comprehension of the methods presented in the following chapters, and will of course enable the reader to go on and start applying these methods to more immediate problems. However, this book is written with the general reader in mind — there is no assumption of programming knowledge, although a few illustrative MATLAB programs appear in the text. The goal is to help the reader understand the logic behind the alternative approaches for forecasting, risk analysis, and decision-making support in volatile financial markets.

Following Wolkenhauer (2001), I struggled to impose a linear ordering on what is essentially a web-like structure. I know my success in this can be only partial. I encourage readers to skip ahead to find more illustrative examples of the concepts raised in earlier parts of the book in succeeding chapters.

I show throughout this book that the application of neural network approximation coupled with evolutionary computational methods for estimation have a predictive edge in out-of-sample forecasting. This predictive edge is relative to standard econometric methods. I do not claim that this predictive edge from neural networks will always lead to opportunities for profitable trading [see Qi (1999)], but any predictive edge certainly enhances the chance of finding such opportunities.

This book grew out of a large and continuing series of lectures given in Latin America, Asia, and Europe, as well as from advanced undergraduate seminars and graduate-level courses at Georgetown University and Boston College. In Latin America, the lectures were first given in São Paulo, Brazil, under the sponsorship of the Brazilian Association of Commercial Bankers (ABBC), in March 1996. These lectures were offered again in March 1997 in São Paulo, in August 1998 at Banco do Brasil in Brasilia, and later that year in Santiago, Chile, at the Universidad Alberto Hurtado.

In Asia and Europe, similar lectures took place at the Monetary Policy and Economic Research Department of Bank Indonesia, under the sponsorship of the United States Agency for International Development, in

January 1996. In May 1997 a further series of lectures on this subject took place under the sponsorship of the Programme for Monetary and Financial Studies of the Department of Economics of the University of Melbourne, and in March of 1998 a similar course was offered at the Facultat d'Economia of the Universitat Ramon Llull sponsored by the Callegi d'Economistes de Calalunya in Barcelona.

The Center for Latin American Economics of the Research Department of the Federal Reserve Bank of Dallas provided the opportunity in the autumn of 1997 to do some of the initial formal research for the financial examples illustrated in this book. In 2003 and early 2004, the Hong Kong Institute of Monetary Research was the center for a summer of research on applications of neural network methods for forecasting deflationary cycles in Hong Kong, and in 2004 the School of Economics and Social Sciences at Singapore Management University and the Institute of Mathematical Sciences at the National University of Singapore were hosts for a seminar and for research on nonlinear principal components

Some of the most useful inputs for the material for this book came from discussions with participants at the International Joint Conference on Neural Networks (IJCNN) meetings in Washington, DC, in 2001, and in Honolulu and Singapore in 2002. These meetings were eye-openers for anyone trained in classical statistics and econometrics and illustrated the breadth of applications of neural network research.

I wish to thank my fellow Jesuits at Georgetown University and in Washington, DC, who have been my "company" since my arrival at Georgetown in 1977, for their encouragement and support in my research undertakings. I also acknowledge my colleagues and students at Georgetown University, as well as economists at the universities, research institutions, and central banks I have visited, for their questions and criticism over the years. We economists are not shy about criticizing one another's work, but for me such criticism has been more gain than pain. I am particularly grateful to the reviewers of earlier versions of this manuscript for Elsevier Academic Press. Their constructive comments gave me new material to pursue and enhanced my own understanding of neural networks.

I dedicate this book to the first member of the latest generation of my clan, Reese Anthony Snyder, born June 18, 2002.

1
Introduction

1.1 Forecasting, Classification, and Dimensionality Reduction

This book shows how neural networks may be put to work for more accurate forecasting, classification, and dimensionality reduction for better decision making in financial markets — particularly in the volatile emerging markets of Asia and Latin America, but also in domestic industrialized-country asset markets and business environments.

The importance of better forecasting, classification methods, and dimensionality reduction methods for better decision making, in the light of increasing financial market volatility and internationalized capital flows, cannot be overexaggerated. The past two decades have witnessed extreme macroeconomic instability, first in Latin America and then in Asia. Thus, both financial analysts and decision makers cannot help but be interested in predicting the underlying *rates of return and spreads,* as well as the *default rates,* in domestic and international credit markets.

With the growth of the market in financial derivatives such as call and put options (which give the right but not the obligation to buy or sell assets at given prices at preset future periods), the pricing of instruments for hedging positions on underlying risky assets and optimal portfolio diversification have become major activities in international investment institutions. One of the key questions facing practitioners in financial markets is the correct pricing of new derivative products as demand for these instruments grows.

To put it bluntly, if practitioners in these markets do not wish to be "taken to the cleaners" by international arbitrageurs and risk management specialists, then they had better learn how to price their derivative offerings in ways that render them arbitrage-free. Correct pricing of risk, of course, crucially depends on the correct understanding of the process driving the underlying rates of return. So correct pricing requires the use of models that give relatively accurate out-of-sample forecasts.

Forecasting simply means understanding which variables lead or help to predict other variables, when many variables interact in volatile markets. This means looking at the past to see what variables are significant leading indicators of the behavior of other variables. It also means a better understanding of the timing of lead–lag relations among many variables, understanding the statistical significance of these lead–lag relationships, and learning which variables are the more important ones to watch as signals for further developments in other returns.

Obviously, if we know the true underlying model generating the data we observe in markets, we will know how to obtain the best forecasts, even though we observe the data with measurement error. More likely, however, the true underlying model may be too complex, or we are not sure which model among many competing ones is the true one. So we have to approximate the true underlying model by approximating models. Once we acknowledge model uncertainty, and that our models are approximations, neural network approaches will emerge as a strong competitor to the standard benchmark linear model.

Classification of different investment or lending opportunities as acceptable or unacceptable risks is a familiar task in any financial or business organization. Organizations would like to be able to discriminate good from bad risks by identifying key characteristics of investment candidates. In a lending environment, a bank would like to identify the likelihood of default on a car loan by readily identifiable characteristics such as salary, years in employment, years in residence, years of education, number of dependents, and existing debt. Similarly, organizations may desire a finer grid for discriminating, from very low, to medium, to very high unacceptable risk, to manage exposure to different types of risk. Neural nets have proven to be very effective classifiers — better than the state-of-the-art methods based on classical statistical methods.[1]

Dimensionality reduction is also a very important component in financial environments. All too often we summarize information about large amounts of data with averages, means, medians, or trimmed means, in which a given

---

[1] Of course, classification has wider applications, especially in the health sciences. For example, neural networks have proven very useful for detection of high or low risks of various forms of cancer, based on information from blood samples and imaging.

percentage of high and low extreme values are eliminated from the sample. The Dow-Jones Industrial Average is simply that: an average price of industrial share prices. Similarly the Standard and Poor 500 is simply the average price of the largest 500 share prices. But averages can be misleading. For example, one student receiving a B grade in all her courses has a B average. Another student may receive A grades in half of his courses and a C grade in the rest. The second student also has a B average, but the performances of the two students are very different. While the grades of the first student cluster around a B grade, the grades of the second student cluster around two grades: an A and a C. It is very important to know if the average reported in the news truly represents where the market is through dimensionality reduction if it is to convey meaningful information.

Forecasting into the future, or *out-of-sample predictions*, as well as classification and dimensionality reduction models, must go beyond diagnostic examination of past data. We use the coefficients obtained from past data to fit new data and make predictions, classification, and dimensionality reduction decisions for the future. As the saying goes, life must be understood looking backwards, but must be lived looking forward. The past is certainly helpful for predicting the future, but we have to know which approximating models to use, in combination with past data, to predict future events. The medium-term strategy of any enterprise depends on the outlook in the coming quarters for both price and quantity developments in its own industry. The success of any strategy depends on how well the forecasts guiding the decision makers work.

Diagnostic and forecasting methods feed back in very direct ways to decision-making environments. Knowing what determines the past, as well as what gives good predictions for the future, gives decision makers better information for making optimal decisions over time. In engineering terms, knowing the underlying "laws of motion" of key variables in a dynamic environment leads to the development of optimal feedback rules. Applying this concept to finance, if the Fed raises the short-term interest rate, how should portfolio managers shift their assets? Knowing how the short-term rates affect a variety of rates of return and how they will affect the future inflation rate can lead to the formulation of a reaction function, in which financial officers shift from risky assets to higher-yield, risk-free assets. We call such a policy function, based on the "laws of motion" of the system, *control*. Business organizations by their nature are interested in diagnostics and prediction so that they may formulate policy functions for effective control of their own future welfare.

Diagnostic examination of past data, forecasting, and control are different activities but are closely related. The policy rule for control, of course, need not be a hard and fast mechanical rule, but simply an operational guide for better decision making. With good diagnostics and forecasting, for example, businesses can better assess the effects of changes in their prices on demand, as well as the likely response of demand to external shocks, and thus how to reset their prices. So it should not be so surprising that good predictive methods are at a premium in research departments for many industries.

Accurate forecasting methods are crucial for portfolio management by commercial and investment banks. Assessing expected returns relative to risk presumes that portfolio strategists understand the distribution of returns. Until recently, most of the control or decision-making analysis has been based on linear dynamic models with normal or log-normal distributions of asset returns. However, finding such a distribution in volatile environments means going beyond simple assumptions of normality or log normality used in conventional models of portfolio strategies. Of course, when we let go of normality, we must get our hands dirty in numerical approximation, and can no longer plug numbers into quick formulae based on normal distributions. But there are clear returns from this extra effort.

The message of this book is that business and financial decision makers now have available the computational power and methods for more accurate diagnostics, forecasting, and control in volatile, increasingly complex, multidimensional environments. Researchers need no longer confine themselves to linear or log-linear models, or assume that underlying stochastic processes are Gaussian or normal in order to obtain forecasts and pinpoint risk–return trade-offs. In short, we can go beyond linearity and normality in our assumptions with the use of neural networks.

1.2 Synergies

The activities of formal diagnostics and forecasting and practical decision making or control in business and finance complement one another, even though mastering each of them requires different types of skills and the exercise or use of different but related algorithms. Applying diagnostic and predictive methods requires knowledge of particular ways to filter or preprocess data for optimum convergence, as well as for estimation, to achieve good diagnostics and out-of-sample accuracy. Decision making in finance, such as buying or selling or setting the pricing of different types of instruments, requires the use of specific assumptions about how to classify risk and about the preferences of investors regarding risk–return trade-offs. Thus, the outcomes crucially depend on the choice of the preference or *welfare index* about acceptable risk and returns over time.

From one perspective, the influence is unidirectional, proceeding from diagnostic and forecasting methods to business and financial decision making. Diagnostics and forecasting simply provide the inputs or stylized facts about expected rates of return and their volatility. These forecasts are the crucial ingredients for pricing decisions, both for firm products and for financial instruments such as call or put options and other more exotic types of derivatives.

From another perspective, however, there may be feedback or bidirectional influence. Knowledge of the objective functions of managers, or their welfare indices, from survey expectations of managers, may be useful leading indicators in forecasting models, particularly in volatile environments. Similarly, the estimated risk, or volatility, derived from forecasting models and the implied risk, given by the pricing decisions of call or put options or swaps in financial markets, may sharply diverge when there is a great deal of uncertainty about the future course of the economy. In both of these cases, the information calculated from survey expectations or from the implied volatilities given by prices of financial derivatives may be used as additional instruments for improving the performance of forecasting models for the underlying rates of return. We may even be interested in predicting the implied volatilities coming from options prices.

Similarly, deciding what price index to use for measuring and forecasting inflation may depend on what the end user of this information intends to do. If the purpose is to help the monetary authority monitor inflationary pressures for setting policy, then price indices that have a great deal of short-term volatility may not be appropriate. In this case, the overly volatile measure of the price level may induce overreactions in the setting of short-term interest rates. By the same token, a price measure that is too smooth may lead to a very passive monetary policy that fails to dampen rising inflationary pressures. Thus, it is useful to distill information from a variety of price indices, or rates of return, to find the movement of the market or the fundamental driving force. This can be done very effectively with neural network approaches.

Unlike hard sciences such as physics or engineering, the measurement and statistical procedures of diagnostics and forecasting are not so cleanly separable from the objectives of the researchers, decision makers, and players in the market. This is a subtle but important point that needs to be emphasized. When we formulate approximating models for the rates of return in financial markets, we are in effect attempting to forecast the forecasts of others. Rates of return rise or fall in reaction to changes in public or private news, because traders are reacting to news and buying or selling assets. Approximating the true underlying model means taking into account, as we formulate our models, how traders — human beings like us — actually learn, process information, and make decisions.

Recent research in macroeconomics by Sargent (1997, 1999), to be discussed in greater detail in the following section, has drawn attention to the fact that the decision makers we wish to approximate with our models are not fully rational, and thus "all-knowing," about their financial environment. Like us, they have to learn what is going on. For this very reason, neural network methods are a natural starting point for approximation in financial markets. Neural networks grew out of the cognitive and brain science disciplines for approximating how information is processed and becomes insight. We illustrate this point in greater detail when we examine the structure of typical neural network frameworks. Suffice it to say, neural network analysis is becoming a key component of the epistemology (philosophy of knowledge) implicit in empirical finance.

1.3   The Interface Problems

The goal of this study is to "break open" the growing literature on neural networks to make the methods accessible, user friendly, and operational for the broader population of economists, analysts, and financial professionals seeking to become more efficient in forecasting. A related goal is to focus the attention of researchers in the fields of neural networks and related disciplines, such as genetic algorithms, to areas in which their tools may have particular advantages over state-of-the-art methods in economics and finance, and thus may make significant contributions to unresolved issues and controversies.

Much of the early development of neural network analysis has been within the disciplines of psychology, neurosciences, and engineering, often related to problems of pattern recognition. Genetic algorithms, which we use for empirically implementing neural networks, have followed a similar pattern of development within applied mathematics, with respect to optimization of dynamic nonlinear and/or discrete systems, moving into the data engineering field.

Thus there is an understandable interface problem for students and professionals whose early formation in economics has been in classical statistics and econometrics. Many of the terms are simply not familiar, or sound odd. For example, a *model* is known as an *architecture*, and we *train* rather than *estimate* a network architecture. A researcher makes use of a *training set* and a *test set* of data, rather than using *in-sample* and *out-of-sample* data. Coefficients are called *weights* and constant terms are *biases*.

Besides these semantic or vocabulary differences, however, many of the applications in the neural network (and broader artificial intelligence) literature simply are not relevant for financial professionals, or if relevant, do not resonate well with the matters at hand. For example, pattern recognition is usually applied to problems of identifying letters of the alphabet for computational translation in linguistics research. A much more interesting example would be to examine recurring patterns such as "bubbles" in high-frequency asset returns data, or the pattern observed in the term structure of interest rates.

Similarly, many of the publications on financial markets by neural network researchers have an ad hoc flavor and do not relate to the broader theoretical infrastructure and fundamental behavioral assumptions used in economics and finance. For this reason, unfortunately, much of this research is not taken seriously by the broader academic community in economics and finance.

The appeal of the neural network approach lies in its assumption of *bounded rationality*: when we forecast in financial markets, we are forecasting the forecasts of others, or approximating the expectations of others. Financial market participants are thus engaged in a learning process, continually adapting prior subjective beliefs from past mistakes.

What makes the neural network approach so appealing in this respect is that it permits threshold responses by economic decision makers to changes in policy or exogenous variables. For example, if the interest rate rises from 3 percent to 3.1 or 3.2 percent, there may be little if any reaction by investors. However, if the interest rate continues to increase, investors will take notice, more and more. If the interest rate crosses a critical threshold, for example, of 5 percent, there may be a massive reaction or "meltdown," with a sell-off of stocks and a rush into government securities.

The basic idea is that reactions of economic decision makers are not linear and proportionate, but asymmetric and nonlinear, to changes in external variables. Neural networks *approximate* this behavior of economic and financial decision making in a very intuitive way.

In this important sense neural networks are different from classical econometric models. In the neural network model, one is not making any specific hypothesis about the values of the coefficients to be estimated in the model, nor, for that matter, any hypothesis about the functional form relating the observed regressor x to an observed output y. Most of the time, we cannot even interpret the meaning of the coefficients estimated in the network, at least in the same way we can interpret estimated coefficients in ordinary econometric models, with a well-defined functional form. In that sense, the neural network differs from the usual econometrics, where considerable effort is made to obtain accurate and consistent, if not unbiased, estimates of particular parameters or coefficients.

Similarly, when nonlinear models are used, too often economists make use of numerical algorithms based on assumptions of continuous or "smooth" data. All too often, these methods break down, or one must make use of repeated estimation, to make sure that the estimates do not represent one of several possible sets of local optimum positions. The use of the genetic algorithm and other evolutionary search algorithms enable researchers to work with discontinuities and to locate with greater probability the global optimum. This is the good news. The bad news is that we have to wait a bit longer to get these results.

The financial sectors of emerging markets, in particular, but also in markets with a great deal of innovation and change, represent a fertile ground for the use of these methods for two reasons, which are interrelated. One is that the data are often very noisy, due either to the thinness of the markets or to the speed with which news becomes dispersed, so that there are obvious asymmetries and nonlinearities that cannot be assumed away. Second, in many instances, the players in these markets are themselves in a process of learning, by trial and error, about policy news or about legal and other changes taking place in the organization of their markets. The parameter estimates of a neural network, by which market participants forecast and make decisions, are themselves the outcome of a learning and search process.

1.4   Plan of the Book

The next chapter takes up the question: What is a neural network? It also takes up the relevance of the "black box criticism" directed against neural network and nonlinear estimation methods. The succeeding chapters ask how we estimate such networks, and then how we evaluate and interpret the results of network estimation.

Chapters 2 through 4 cover the basic theory of neural networks. These chapters, by far, are the most technical chapters of the book. They are oriented to people familiar with classical statistics and linear regression. The goal is to relate recent developments in the neural network and related genetic search literature to the way econometricians routinely do business, particularly with respect to the linear autoregressive model. It is intended as a refresher course for those who wish to review their econometrics. However, in succeeding chapters we flesh out with specific data sets the more technical points developed here. The less technically oriented reader may skim through these chapters at the first reading and then return to them as a cross-reference periodically, to clarify definitions of alternative procedures reported with the examples of later chapters.

These chapters contrast the setup of the neural network with the standard linear model. While we do not elaborate on the different methods for estimating linear autoregressive models, since these topics are extensively covered in many textbooks on econometrics, there is a detailed treatment of the nonlinear estimation process for neural networks. We also lay out the basics of genetic algorithms as well as with more familiar gradient or quasi-Newtonian methods based on the calculation of first- and second-order derivatives for estimating the neural network models. Evolutionary computation involves coupling the global genetic search methods with local gradient methods.

Chapters 3 and 4, on estimation and evaluation, also review the basic metrics or statistical tests we use to evaluate the success of a model, whether the model is the standard linear one or a nonlinear neural network. We also treat the ways we need to filter, adjust, or preprocess data prior to statistical estimation and evaluation. It should be clear from this chapter that the straw man or benchmark of this book is the standard linear or linear autoregressive model. Throughout the chapters, the criteria for success of neural network forecasting is measured relative to the standard linear model.

The fifth chapter presents several applications for evaluating the performance of alternative networks with artificial data to illustrate the points made in the previous three chapters. The reason for using artificial data is that we can easily verify the accuracy of the network model, relative to other approaches, if we know the true model generating the data. This chapter shows, in one example, how artificial data generated with the Black-Scholes option pricing model, as well as with more advanced option pricing formulae, may be closely matched, out of sample, by a neural network. Thus, the neural network may be used to complement more complicated options or derivative pricing models for setting the initial market price of such instruments. This section shows very clearly the relative accuracy or predictive power of the neural network or genetic algorithm.

Following an application to artificial data, we apply, in Chapter 6, neural network methods to actual forecasting problems: at the industrial level, in the quantity of automobiles as a function of the price index as well as aggregate interest rates and disposable income; at the financial level, predicting spreads in corporate bonds (relative to 10-year U.S. Treasury bonds) as a function of default rates, the real exchange rate, industrial production, the share-market index, and indices of market expectations. The seventh chapter examines inflation and deflation forecasting at the macroeconomic level, with sample data from Hong Kong and Japan. Chapter 8 takes up classification problems, specifically credit card default and banking intervention, as functions of observed characteristics, using both categorical and more familiar continuous variables as the inputs. Chapter 9 shows the usefulness of neural networks for distilling information from market volatilities, for obtaining an overall sense of market volatility and with nonlinear principal components, and evaluates the performance of this method relative to linear principal component analysis.

While time-series analysis, classification, and dimensionality reduction are taken up as separate tasks, frequently they can be synergistic. For example, dimensionality reduction can be used to reduce the number of regressors in a model for forecasting. Similarly, the forecasts of a time-series model, representing expectations of inflation or future growth, may be inputs at any given time in a classification model. Time-series forecasting, classification, and dimensionality reduction are very useful for understanding a wide variety of financial market issues.

Each of the chapters concludes not only with a short summary, but also with discussion questions, references to MATLAB programs available on the website, and suggestions for further exercises. The programs are written especially for this book. Certainly they are not meant to be examples of efficient programming code. There is the ever-present trade-off between transparency and efficiency in writing programming code. My first goal in writing these programs was to make the programs "transparent" to myself! Readers are invited to change, amend, and mutate these programs to make them even more efficient and transparent for themselves. These MATLAB programs require the optimization and statistics toolbox. We also make use of the symbolic toolbox for a few exercises.

There is much more that could be part of this book. There is no discussion, in particular, of estimation and forecasting with intra-daily or real-time data. This is a major focus of recent financial market research, particularly the new micro-structure exchange-rate economics. One reason for bypassing the use of real-time data is that it is usually proprietary. While estimation results can be reported in scholarly research, the data sets, without special arrangements, cannot be made available to other researchers for replication and further study. In this study, we want to encourage the readers to use both the data sets and MATLAB programs of this book to enhance their own learning. For this reason, we stay with familiar examples as the best way to illustrate the predictive power that comes from harnessing neural networks with evolutionary computation.

Similarly, there is no discussion of forecasting stock-market returns or the rates of change of other asset prices or exchange rates. While many researchers have tried to show the profitable use of trading strategies based on neural network out-of-sample forecasting relative to other strategies [Qi (1999)], a greater payoff of neural networks in financial markets may come from volatility forecasting.

Part I

Econometric Foundations

2

What Are Neural Networks?

2.1 Linear Regression Model

The rationale for the use of the neural network is forecasting or predicting a given target or output variable y from information on a set of observed input variables x. In time series, the set of input variables x may include lagged variables, the current variables of x, and lagged values of y. In forecasting, we usually start with the linear regression model, given by the following equation:

$$y_t = \sum \beta_k x_{k,t} + \epsilon_t \qquad (2.1a)$$

$$\epsilon_t \sim N(0, \sigma^2) \qquad (2.1b)$$

where the variable ϵ_t is a random disturbance term, usually assumed to be normally distributed with mean zero and constant variance σ^2, and $\{\beta_k\}$ represents the parameters to be estimated. The set of estimated parameters is denoted $\{\widehat{\beta}_k\}$, while the set of forecasts of y generated by the model with the coefficient set $\{\widehat{\beta}_k\}$ is denoted by $\{\widehat{y}_t\}$. The goal is to select $\{\widehat{\beta}_k\}$ to minimize the sum of squared differences between the actual observations y and the observations predicted by the linear model, $\widehat{y}$.

In time series, the input and output variables, $[y \; x]$, have subscript t, denoting the particular observation date, with the earliest observation starting at $t = 1$.[1] In the standard econometrics courses, there are a variety of methods for estimating the parameter set $\{\beta_k\}$, under a variety of alternative assumptions about the distribution of the disturbance term, ϵ_t, about the constancy of its variance, σ^2, as well as about the independence of the distribution of the input variables x_k with respect to the disturbance term, ϵ_t.

The goal of the estimation process is to find a set of parameters for the regression model, given by $\{\widehat{\beta}_k\}$, to minimize Ψ, defined as the sum of squared differences, or residuals, between the observed or target or output variable y and the model-generated variable $\widehat{y}$, over all the observations. The estimation problem is posed in the following way:

$$\underset{\beta}{Min}\Psi = \sum_{t=1}^{T}\widehat{\epsilon}_t^2 = \sum_{t=1}^{T}(y_t - \widehat{y}_t)^2 \tag{2.2}$$

$$\text{s.t.} \quad y_t = \sum \beta_k x_{k,t} + \epsilon_t \tag{2.3}$$

$$\widehat{y}_t = \sum \widehat{\beta}_k x_{k,t} \tag{2.4}$$

$$\epsilon_t \sim N(0, \sigma^2) \tag{2.5}$$

A commonly used linear model for forecasting is the autoregressive model:

$$y_t = \sum_{i=1}^{k*} \beta_i y_{t-i} + \sum_{j=1}^{k} \gamma_j x_{j,t} + \epsilon_t \tag{2.6}$$

in which there are k independent x variables, with coefficient γ_j for each x_j, and k^* lags for the dependent variable y, with, of course $k + k^*$ parameters, $\{\beta\}$ and $\{\gamma\}$, to estimate. Thus, the longer the lag structure, the larger the number of parameters to estimate and the smaller the degrees of freedom of the overall regression estimates.[2]

The number of output variables, of course, may be more than one. But in the benchmark linear model, one may estimate and forecast each output variable $y_j, j = 1, \ldots, j^*$ with a series of J^* independent linear models. For j^* output or dependent variables, we estimate $(J^* \cdot K)$ parameters.

---

[1] In cross-section analysis, the subscript for $[y\ x]$ can be denoted by an identifier i, which refers to the particular individuals, households, or other economic entities being examined. In cross-section analysis, the ordering of the observations with respect to particular observations does not matter.

[2] In the time-series model this model is known as the linear ARX model, since there are autoregressive components, given by the lagged y variables, as well as exogenous x variables.

The linear model has the useful property of having a closed-form solution for solving the estimation problem, which minimizes the sum of squared differences between y and $\widehat{y}$. The solution method is known as linear regression. It has the advantage of being very quick. For short-run forecasting, the linear model is a reasonable starting point, or benchmark, since in many markets one observes only small symmetric changes in the variable to be predicted around a long-term trend. However, this method may not be especially accurate for volatile financial markets. There may be nonlinear processes in the data. Slow upward movements in asset prices followed by sudden collapses, known as bubbles, are rather common. Thus, the linear model may fail to capture or forecast well sharp turning points in data. For this reason, we turn to nonlinear forecasting techniques.

2.2   GARCH Nonlinear Models

Obviously, there are many types of nonlinear functional forms to use as an alternative to the linear model. Many nonlinear models attempt to capture the true or underlying nonlinear processes through parametric assumptions with specific nonlinear functional forms. One popular example of this approach is the GARCH-In-Mean or GARCH-M model.[3] In this approach, the variance of the disturbance term directly affects the mean of the dependent variable and evolves through time as a function of its own past value and the past squared prediction error. For this reason, the time-varying variance is called the *conditional variance*. The following equations describe a typical parametric GARCH-M model:

$$\sigma_t^2 = \delta_0 + \delta_1 \sigma_{t-1}^2 + \delta_2 \epsilon_{t-1}^2 \tag{2.7}$$

$$\epsilon_t \approx \phi(0, \sigma_t^2) \tag{2.8}$$

$$y_t = \alpha + \beta \sigma_t + \epsilon_t \tag{2.9}$$

where y is the rate of return on an asset, α is the expected rate of appreciation, and ϵ_t is the normally distributed disturbance term, with mean zero and conditional variance σ_t^2, given by $\phi(0, \sigma_t^2)$. The parameter β represents the risk premium effect on the asset return, while the parameters δ_0, δ_1, and δ_2 define the evolution of the conditional variance. The risk premium reflects the fact that investors require higher returns to take on higher risks in a market. We thus expect $\beta > 0$.

---

[3]GARCH stands for generalized autoregresssive conditional heteroskedasticity, and was introduced by Bollerslev (1986, 1987) and Engle (1982). Engle received the Nobel Prize in 2003 for his work on this model.

The GARCH-M model is a stochastic recursive system, given the initial conditions σ_0^2 and ϵ_0^2, as well as the estimates for $\alpha, \beta, \delta_0, \delta_1$, and δ_2. Once the conditional variance is given, the random shock is drawn from the normal distribution, and the asset return is fully determined as a function of its own mean, the random shock, and the risk premium effect, determined by $\beta \sigma_t$.

Since the distribution of the shock is normal, we can use maximum likelihood estimation to come up with estimates for $\alpha, \beta, \delta_0, \delta_1$, and δ_2. The likelihood function L is the joint probability function for $\widehat{y}_t = y_t$, for $t = 1, \ldots, T$. For the GARCH-M models, the likelihood function has the following form:

$$L_t = \prod_{t=1}^{T} \sqrt{\frac{1}{2\pi \widehat{\sigma}_t^2}} \exp\left[-\frac{(y_t - \widehat{y}_t)^2}{2\widehat{\sigma}_t^2}\right] \tag{2.10}$$

$$\widehat{y}_t = \widehat{\alpha} + \widehat{\beta}\widehat{\sigma}_t \tag{2.11}$$

$$\widehat{\epsilon}_t = y_t - \widehat{y}_t \tag{2.12}$$

$$\widehat{\sigma}_t^2 = \widehat{\delta}_0 + \widehat{\delta}_1 \widehat{\sigma}_{t-1}^2 + \widehat{\delta}_2 \widehat{\epsilon}_{t-1}^2 \tag{2.13}$$

where the symbols $\widehat{\alpha}, \widehat{\beta}, \widehat{\delta}_0, \widehat{\delta}_1$, and $\widehat{\delta}_2$ are the estimates of the underlying parameters, and Π is the multiplication operator, $\Pi_{i=1}^{2} x_i = x_1 \cdot x_2$. The usual method for obtaining the parameter estimates maximizes the sum of the *logarithm* of the likelihood function, or log-likelihood function, over the entire sample T, from $t = 1$ to $t = T$, with respect to the choice of coefficient estimates, subject to the restriction that the variance is greater than zero, given the initial condition $\widehat{\sigma}_0^2$ and $\widehat{\epsilon}_{t-1}^2$:[4]

$$\underset{\{\widehat{\alpha},\widehat{\beta},\widehat{\delta}_0,\widehat{\delta}_1,\widehat{\delta}_2\}}{Max} \sum_{t=1}^{T} \ln(L_t) = \sum_{t=1}^{T}\left(-.5\ln(2\pi) - .5\ln(\widehat{\sigma}_t) - .5\left[\frac{(y_t - \widehat{y}_t)^2}{\widehat{\sigma}_t^2}\right]\right) \tag{2.14}$$

$$\text{s.t.} \quad : \quad \widehat{\sigma}_t^2 > 0, t = 1, 2, \ldots, T \tag{2.15}$$

The appeal of the GARCH-M approach is that it pins down the source of the nonlinearity in the process. The conditional variance is a nonlinear transformation of past values, in the same way that the variance measure

---

[4]Taking the sum of the logarithm of the likelihood function produces the same estimates as taking the product of the likelihood function, over the sample, from $t = 1, 2, \ldots, T$.

is a nonlinear transformation of past prediction errors. The justification of using conditional variance as a variable affecting the dependent variable is that conditional variance represents a well-understood risk factor that raises the required rate of return when we are forecasting asset price dynamics.

One of the major drawbacks of the GARCH-M method is that minimization of the log-likelihood functions is often very difficult to achieve. Specifically, if we are interested in evaluating the statistical significance of the coefficient estimates, $\widehat{\alpha}, \widehat{\beta}, \widehat{\delta}_0, \widehat{\delta}_1$, and $\widehat{\delta}_2$, we may find it difficult to obtain estimates of the confidence intervals. All of these difficulties are common to maximum likelihood approaches to parameter estimation.

The parametric GARCH-M approach to the specification of nonlinear processes is thus restrictive: we have a specific set of parameters we want to estimate, which have a well-defined meaning, interpretation, and rationale. We even know how to estimate the parameters, even if there is some difficulty. The good news of GARCH-M models is that they capture a well-observed phenomenon in financial time series, that periods of high volatility are followed by high volatility and periods of low volatility are followed by similar periods.

However, the restrictiveness of the GARCH-M approach is also its drawback: we are limited to a well-defined set of parameters, a well-defined distribution, a specific nonlinear functional form, and an estimation method that does not always converge to parameter estimates that make sense. With specific nonlinear models, we thus lack the flexibility to capture alternative nonlinear processes.

2.2.1 *Polynomial Approximation*

With neural network and other approximation methods, we approximate an unknown nonlinear process with less-restrictive semi-parametric models. With a polynomial or neural network model, the functional forms are given, but the degree of the polynomial or the number of neurons are not. Thus, the parameters are neither limited in number, nor do they have a straightforward interpretation, as the parameters do in linear or GARCH-M models. For this reason, we refer to these models as *semi-parametric*. While GARCH and GARCH-M models are popular models for nonlinear financial econometrics, we show in Chapter 3 how well a rather simple neural network approximates a time series that is generated by a calibrated GARCH-M model.

The most commonly used approximation method is the polynomial expansion. From the Weierstrass Theorem, a polynomial expansion around a set of inputs x with a progressively larger power P is capable of approximating to a given degree of precision any unknown but continuous function

$y = g(x).$[5] Consider, for example, a second-degree polynomial approximation of three variables, $[x_{1t}, x_{2t}, x_{3t}]$, where g is unknown but assumed to be
a continuous function of arguments x_1, x_2, x_3. The approximation formula
becomes:

$$y_t = \beta_0 + \beta_1 x_{1t} + \beta_2 x_{2t} + \beta_3 x_{3t} + \beta_4 x_{1t}^2 + \beta_5 x_{2t}^2 + \beta_6 x_{3t}^2 + \beta_7 x_{1t} x_{2t}$$
$$+ \beta_8 x_{2t} x_{3t} + \beta_9 x_{1t} x_{3t} \tag{2.16}$$

Note that the second-degree polynomial approximation with three arguments or dimensions has three cross-terms, with coefficients given by
$\{\beta_7, \beta_8, \beta_9\}$, and requires ten parameters. For a model of several arguments,
the number of parameters rises exponentially with the degree of the polynomial expansion. This phenomenon is known as the *curse of dimensionality*
in nonlinear approximation. The price we have to pay for an increasing
degree of accuracy is an increasing number of parameters to estimate, and
thus a decreasing number of degrees of freedom for the underlying statistical
estimates.

2.2.2   Orthogonal Polynomials

Judd (1999) discusses a wider class of polynomial approximators, called
orthogonal polynomials. Unlike the typical polynomial based on raising the
variable x to powers of higher order, these classes of polynomials are based
on sine, cosine, or alternative exponential transformations of the variable
x. They have proven to be more efficient approximators than the power
polynomial.

Before making use of these orthogonal polynomials, we must transform
all of the variables $[y, x]$ into the interval $[-1, 1]$. For any variable x, the
transformation to a variable x^* is given by the following formula:

$$x^* = \frac{2x}{\max(x) - \min(x)} - \frac{\min(x) + \max(x)}{\max(x) - \min(x)} \tag{2.17}$$

The exact formulae for these orthogonal polynomials are complicated [see
Judd (1998), p. 204, Table 6.3]. However, these polynomial approximators
can be represented rather easily in a recursive manner. The Tchebeycheff

---

[5]See Miller, Sutton, and Werbos (1990), p. 118.

polynomial expansion $T(x^*)$ for a variable x^* is given by the following recursive system:[6]

$$T_0(x^*) = 1$$

$$T_1(x^*) = x^*$$

$$T_{i+1}(x^*) = 2x^* T_i(x^*) - T_{i-1}(x^*) \qquad (2.18)$$

The Hermite expansion $H(x^*)$ is given by the following recursive equations:

$$H_0(x^*) = 1$$

$$H_1(x^*) = 2x^*$$

$$H_{i+1}(x^*) = 2x^* H_i(x^*) - 2i H_{i-1}(x^*) \qquad (2.19)$$

The Legendre expansion $L(x^*)$ has the following form:

$$L_0(x^*) = 1$$

$$L_1(x^*) = 1 - x^*$$

$$L_{i+1}(x^*) = \left(\frac{2i+1}{i+1}\right) L_i(x^*) - \frac{i}{i+1} L_{i-1}(x^*) \qquad (2.20)$$

Finally, the Laguerre expansion $LG(x^*)$ is represented as follows:

$$LG_0(x^*) = 1$$

$$LG_1(x^*) = 1 - x^*$$

$$LG_i(x^*) = \left(\frac{2i+1-x^*}{i+1}\right) LG_i(x^*) - \frac{i}{i+1} LG_{i-1}(x^*) \qquad (2.21)$$

Once these polynomial expansions are obtained for a given variable x^*, we simply approximate y^* with a linear regression. For two variables, $[x_1, x_2]$ with expansion $P1$ and $P2$ respectively, the approximation is given by the following expression:

$$y_t^* = \sum_{i=1}^{P1} \sum_{j=1}^{P2} \beta_{ij} T_i(x_{1t}^*) T_j(x_{2t}^*) \qquad (2.22)$$

---

[6]There is a long-standing controversy about the proper spelling of the first polynomial. Judd refers to the *Tchebeycheff* polynomial, whereas Heer and Maussner (2004) write about the *Chebeyshev* polynomal.

To retransform a variable y^* back into the interval $[\min(y), \max(y)]$, we use the following expression:

$$y = \frac{(y^* + 1)[\max(y) - \min(y)]}{2} + \min(y)$$

The network is an alternative to the parametric linear, GARCH-M models, and semi-parametric polynomial approaches for approximating a nonlinear system. The reason we turn to the neural network is simple and straightforward. The goal is to find an approach or method that forecasts well data generated by often unknown and highly nonlinear processes, with as few parameters as possible, and which is easier to estimate than parametric nonlinear models. Succeeding chapters show that the neural network approach does this better — in terms of accuracy and parsimony — than the linear approach. The network is as accurate as the polynomial approximations with fewer parameters, or more accurate with the same number of parameters. It is also much less restrictive than the GARCH-M models.

2.3 Model Typology

To locate the neural network model among different types of models, we can differentiate between *parametric* and *semi-parametric* models, and models that have and do not have *closed-form* solutions. The typology appears in Table 2.1.

Both linear and polynomial models have closed-form solutions for estimation of the regression coefficients. For example, in the linear model $y = x\beta$, written in matrix form, the typical ordinary least squares (OLS) estimator is given by $\widehat{\beta} = (x'x)^{-1}x'y$. The coefficient vector $\widehat{\beta}$ is a simple linear function of the variables $[y \; x]$. There is no problem of convergence or multiple solutions: once we know the variable set $[y \; x]$, we know the estimator of the coefficient vector, $\widehat{\beta}$. For a polynomial model, in which the dependent variable y is a function of higher powers of the regressors x, the coefficient vector is calculated in the same way as OLS. We simply redefine the regressors in terms of a matrix z, representing polynomial

TABLE 2.1. Model Typology

Closed-Form Solution	Parametric	Semi-Parametric
Yes	Linear	Polynomial
No	GARCH-M	Neural Network

expansions of the regressors x, and calculate the polynomial coefficient vector as $\widehat{\beta} = (z'z)^{-1}z'y$.

Both the GARCH-M and the neural network models are examples of models that do not have closed-form solutions for the coefficient vector of the respective model. We discuss many of the methods for obtaining solutions for the coefficient vector for these models in the following sections. What is clear from Table 2.1, moreover, is that we have a clear-cut choice between linear and neural network models. The linear model may be a very imprecise approximation to the real world, but it gives very easy, quick, exact solutions. The neural network may be a more precise approximation, capturing nonlinear behavior, but it does not have exact, easy-to-obtain solutions. Without a closed-form solution, we have to use approximate solutions. In fact, as Michalewicz and Fogel (2002) point out, this polarity reflects the difficulties in problem solving in general. It is difficult to obtain good solutions to important problems, either because we have to use an imprecise model approximation (such as a linear model) which has an exact solution, or we have to use an approximate solution for a more precise, complex model approximation [Michalewicz and Fogel (2002), p. 19].

2.4   What Is A Neural Network?

Like the linear and polynomial approximation methods, a neural network relates a set of input variables $\{x_i\}, i = 1, \ldots, k$, to a set of one or more output variables, $\{y_j\}, j = 1, \ldots, k*$. The difference between a neural network and the other approximation methods is that the neural network makes use of one or more hidden layers, in which the input variables are squashed or transformed by a special function, known as a logistic or logsigmoid transformation. While this hidden layer approach may seem esoteric, it represents a very efficient way to model nonlinear statistical processes.

2.4.1   Feedforward Networks

Figure 2.1 illustrates the architecture on a neural network with one hidden layer containing two neurons, three input variables $\{x_i.\}, i = 1, 2, 3$, and one output y.

We see *parallel processing*. In addition to the sequential processing of typical linear systems, in which only observed inputs are used to predict an observed output by weighting the input neurons, the two neurons in the hidden layer process the inputs in a parallel fashion to improve the predictions. The connectors between the input variables, often called *input neurons*, and the neurons in the hidden layer, as well as the connectors between the hidden-layer neurons and the output variable, or *output neuron*, are

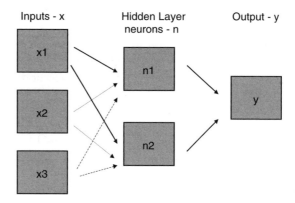

FIGURE 2.1. Feedforward neural network

called *synapses*.[7] Most problems we work with, fortunately, do not involve a large number of neurons engaging in parallel processing, thus the *parallel processing advantage*, which applies to the way the brain works with its massive number of neurons, is not a major issue.

This single-layer feedforward or multiperceptron network with one hidden layer is the most basic and commonly used neural network in economic and financial applications. More generally, the network represents the way the human brain processes input sensory data, received as input neurons, into recognition as an output neuron. As the brain develops, more and more neurons are interconnected by more synapses, and the signals of the different neurons, working in parallel fashion, in more and more hidden layers, are combined by the synapses to produce more nuanced insight and reaction.

Of course, very simple input sensory data, such as the experience of heat or cold, need not lead to processing by very many neurons in multiple hidden layers to produce the recognition or insight that it is time to turn up the heat or turn on the air conditioner. But as experiences of input sensory data become more complex or diverse, more hidden neurons are activated, and insight as well as decision is a result of proper weighting or combining signals from many neurons, perhaps in many hidden layers.

A commonly used application of this type of network is in pattern recognition in neural linguistics, in which handwritten letters of the alphabet are decoded or interpreted by networks for machine translation. However, in

---

[7]The linear model, of course, is a special case of the feedforward network. In this case, the one neuron in the hidden layer is a linear activation function which connects to the one output layer with a weight on unity.

economic and financial applications, the combining of the input variables into various neurons in the hidden layer has another interpretation. Quite often we refer to latent variables, such as expectations, as important driving forces in markets and the economy as a whole. Keynes referred quite often to "animal spirits" of investors in times of boom and bust, and we often refer to bullish (optimistic) or bearish (pessimistic) markets. While it is often possible to obtain survey data of expectations at regular frequencies, such survey data come with a time delay. There is also the problem that how respondents reply in surveys may not always reflect their true expectations.

In this context, the meaning of the hidden layer of different interconnected processing of sensory or observed input data is simple and straightforward. Current and lagged values of interest rates, exchange rates, changes in GDP, and other types of economic and financial news affect further developments in the economy by the way they affect the underlying subjective expectations of participants in economic and financial markets. These subjective expectations are formed by human beings, using their brains, which store memories coming from experiences, education, culture, and other models. All of these interconnected neurons generate expectations or forecasts which lead to reactions and decisions in markets, in which people raise or lower prices, buy or sell, and act bullishly or bearishly. Basically, actions come from forecasts based on the parallel processing of interconnected neurons.

The use of the neural network to model the process of decision making is based on the *principle of functional segregation*, which Rustichini, Dickhaut, Ghirardato, Smith, and Pardo (2002) define as stating that "not all functions of the brain are performed by the brain as a whole" [Rustichini et al. (2002), p. 3]. A second principle, called the *principle of functional integration*, states that "different networks of regions (of the brain) are activated for different functions, with overlaps over the regions used in different networks" [Rustichini et al. (2002), p. 3].

Making use of experimental data and brain imaging, Rustichini, Dickhaut, Ghirardato, Smith, and Pardo (2002) offer evidence that subjects make decisions based on approximations, particularly when subjects act with a short response time. They argue for the existence of a "specialization for processing approximate numerical quantities" [Rustichini et al. (2002), p. 16].

In a more general statistical framework, neural network approximation is a *sieve estimator*. In the univariate case, with one input x, an approximating function of order m, Ψ_m, is based on a non-nested sequence of approximating spaces:

$$\Psi_m = [\psi_{m,0}(x), \psi_{m,1}(x), \ldots \psi_{m,m}(x)] \qquad (2.23)$$

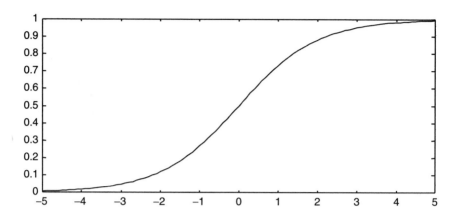

FIGURE 2.2. Logsigmoid function

Beresteanu (2003) points out that each finite expansion, $\psi_{m,0}(x), \psi_{m,1}(x),$ $\ldots \psi_{m,m}(x)$, can potentially be based on a different set of functions [Beresteanu (2003), p. 9]. We now discuss the most commonly used functional forms in the neural network literature.

2.4.2   Squasher Functions

The neurons process the input data in two ways: first by forming linear combinations of the input data and then by "squashing" these linear combinations through the logsigmoid function. Figure 2.2 illustrates the operation of the typical logistic or logsigmoid activation function, also known as a squasher function, on a series ranging from -5 to $+5$. The inputs are thus transformed by the squashers before transmitting their effects on the output.

The appeal of the logsigmoid transform function comes from its threshold behavior, which characterizes many types of economic responses to changes in fundamental variables. For example, if interest rates are already very low or very high, small changes in this rate will have very little effect on the decision to purchase an automobile or other consumer durable. However, within critical ranges between these two extremes, small changes may signal significant upward or downward movements and therefore create a pronounced impact on automobile demand.

Furthermore, the shape of the logsigmoid function reflects a form of learning behavior. Often used to characterize learning by doing, the function becomes increasingly steep until some inflection point. Thereafter the function becomes increasingly flat and its slope moves exponentially to zero.

Following the same example, as interest rates begin to increase from low levels, consumers will judge the probability of a sharp uptick or downtick in the interest rate based on the currently advertised financing packages. The more experience they have, up to some level, the more apt they are to interpret this signal as the time to take advantage of the current interest rate, or the time to postpone a purchase. The results are markedly different from those experienced at other points on the temporal history of interest rates. Thus, the nonlinear logsigmoid function captures a threshold response characterizing bounded rationality or a learning process in the formation of expectations.

Kuan and White (1994) describe this threshold feature as the fundamental characteristic of nonlinear response in the neural network paradigm. They describe it as the "tendency of certain types of neurons to be quiescent of modest levels of input activity, and to become active only after the input activity passes a certain threshold, while beyond this, increases in input activity have little further effect" [Kuan and White (1994), p. 2].

The following equations describe this network:

$$n_{k,t} = \omega_{k,0} + \sum_{i=1}^{i^*} \omega_{k,i} x_{i,t} \tag{2.24}$$

$$N_{k,t} = L(n_{k,t}) \tag{2.25}$$

$$= \frac{1}{1 + e^{-n_{k,t}}} \tag{2.26}$$

$$y_t = \gamma_0 + \sum_{k=1}^{k^*} \gamma_k N_{k,t} \tag{2.27}$$

where $L(n_{k,t})$ represents the logsigmoid activation function with the form $\frac{1}{1+e^{-n_{k,t}}}$. In this system there are i^* input variables $\{x\}$, and k^* neurons. A linear combination of these input variables observed at time t, $\{x_{i,t}\}$, $i = 1, \ldots, i^*$, with the coefficient vector or set of input weights $\omega_{k,i}$, $i = 1, \ldots, i^*$, as well as the constant term, $\omega_{k,0}$, form the variable $n_{k,t}$. This variable is squashed by the logistic function, and becomes a neuron $N_{k,t}$ at time or observation t. The set of k^* neurons at time or observation index t are combined in a linear way with the coefficient vector $\{\gamma_k\}$, $k = 1, \ldots, k^*$, and taken with a constant term γ_0, to form the forecast $\hat{y}_t$ at time t. The feedforward network coupled with the logsigmoid activation functions is also known as the *multi-layer perception* or *MLP* network. It is the basic workhorse of the neural network forecasting approach, in the sense that researchers usually start with this network as the first representative network alternative to the linear forecasting model.

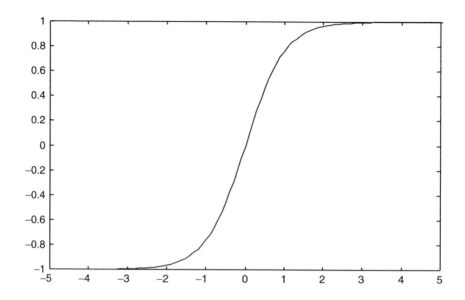

FIGURE 2.3. Tansig function

An alternative activation function for the neurons in a neural network is the hyperbolic tangent function. It is also known as the *tansig* or *tanh* function. It squashes the linear combinations of the inputs within the interval $[-1, 1]$, rather than $[0, 1]$ in the logsigmoid function. Figure 2.3 shows the behavior of this alternative function.

The mathematical representation of the feedforward network with the tansig activation function is given by the following system:

$$n_{k,t} = \omega_{k,0} + \sum_{i=1}^{i^*} \omega_{k,i} x_{i,t} \tag{2.28}$$

$$N_{k,t} = T(n_{k,t}) \tag{2.29}$$

$$= \frac{e^{n_{k,t}} - e^{-n_{k,t}}}{e^{n_{k,t}} + e^{-n_{k,t}}} \tag{2.30}$$

$$y_t = \gamma_0 + \sum_{k=1}^{k^*} \gamma_k N_{k,t} \tag{2.31}$$

where $T(n_{k,t})$ is the tansig activation function for the input neuron $n_{k,t}$.

Another commonly used activation function for the network is the familiar cumulative Gaussian function, commonly known to statisticians as the

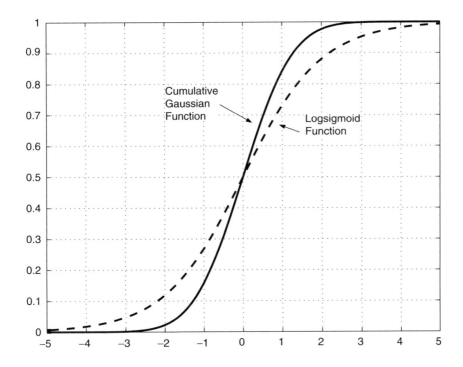

FIGURE 2.4. Gaussian function

normal function. Figure 2.4 pictures this function as well as the logsigmoid function.

The Gaussian function does not have as wide a distribution as the logsigmoid function, in that it shows little or no response when the inputs take extreme values (below -2 or above $+2$ in this case), whereas the logsigmod does show some response. Moreover, within critical changes, such as $[-2, 0]$ and $[0, 2]$, the slope of the cumulative Gaussian function is much steeper. The mathematical representation of the feedforward network with the Gaussian activation functions is given by the following system:

$$n_{k,t} = \omega_{k,0} + \sum_{i=1}^{i^*} \omega_{k,i} x_{i,t} \tag{2.32}$$

$$N_{k,t} = \Phi(n_{k,t}) \tag{2.33}$$

$$= \int_{-\infty}^{n_{k,t}} \sqrt{\frac{1}{2\pi}} e^{-.5n_{k,t}^2} \tag{2.34}$$

$$y_t = \gamma_0 + \sum_{k=1}^{k^*} \gamma_k N_{k,t} \qquad (2.35)$$

where $\Phi(n_{k,t})$ is the standard cumulative Gaussian function.[8]

2.4.3 Radial Basis Functions

The radial basis network function (RBF) network makes use of the radial basis or Gaussian density function as the activation function, but the structure of the network is different from the feedforward or MLP networks we have discussed so far. The input neuron may be a linear combination of regressors, as in the other networks, but there is only one input signal, only one set of coefficients of the input variables x. The signal from this input layer is the same to all the neurons, which in turn are Gaussian transformations, around k^* different means, of the input signals. Thus the input signals have different centers for the radial bases or normal distributions. The differing Gaussian transformations are combined in a linear fashion for forecasting the output.

The following system describes a radial basis network:

$$\underset{<\omega,\mu,\gamma>}{Min} \sum_{t=0}^{T} (y_t - \widehat{y}_t)^2 \qquad (2.36)$$

$$n_t = \omega_0 + \sum_{i=1}^{i^*} \omega_i x_{i,t} \qquad (2.37)$$

$$R_{k,t} = \phi(n_t; \boldsymbol{\mu}_k) \qquad (2.38)$$

$$= \frac{1}{\sqrt{2\pi\sigma_{n-\mu_k}}} \exp\left(\frac{-[n_t - \boldsymbol{\mu}_k]}{\sigma_{n-\mu_k}} \right)^2 \qquad (2.39)$$

$$\widehat{y}_t = \gamma_0 + \sum_{k=1}^{k^*} \gamma_k N_{k,t} \qquad (2.40)$$

where x again represents the set of input variables and n represents the linear transformation of the input variables, based on weights ω. We choose k^* different centers for the radial basis transformation, $\boldsymbol{\mu}_k, k = 1, \ldots, k^*$, calculate the k^* standard error implied by the different centers, $\boldsymbol{\mu}_k$, and

---

[8]The Gaussian function, used as an activation function in a multilayer perceptron or feedforward network, is not a radial basis function network. We discuss that function next.

obtain the k^* different radial basis functions, R_k. These functions in turn are combined linearly to forecast y with weights γ (which include a constant term). Optimizing the radial basis network involves choosing the coefficient sets $\{\omega\}$ and $\{\gamma\}$ as well as the k^* centers of radial basis functions $\{\mu\}$.

Haykin (1994) points out a number of important differences between the RBF and the typical multilayer perceptron network; we note two. First, the RBF network has at most one hidden layer, whereas an MLP network may have many (though in practice we usually stay with one hidden layer). Second, the activation function of the RBF network computes the Euclidean norm or distance (based on the Gaussian transformation) between the signal from the input vector and the center of that unit, whereas the MLP or feedforward network computes the inner products of the inputs and the weights for that unit.

Mandic and Chambers (2001) point out that both the feedforward or multilayer perceptron networks and radial basis networks have good approximation properties, but they note that "an MLP network can always simulate a Gaussian RBF network, whereas the converse is true only for certain values of the bias parameter" [Mandic and Chambers (2001), p. 60].

2.4.4 Ridgelet Networks

Chen, Racine, and Swanson (2001) have shown the ridgelet function to be a useful and less-restrictive alternative to the Gaussian activation functions used in the "radial basis" type sieve network. Such a function, denoted by $R(\cdot)$, can be chosen for a suitable value of m as $\nabla^{m-1}\phi$, where ∇ represents the gradient operator and ϕ is the standard Gaussian density function. Setting $m = 6$, the ridgelet function is defined in the following way:

$$R(x) = \nabla^{m-1}\phi$$

$$m = 6 \implies R(x) = \left(-15x + 10x^3 - x^5\right)\exp\left(-.5x^2\right)$$

The curvature of this function, for the same range of input values, appears in Figure 2.5.

The ridgelet function, like the Gaussian density function, has very low values for the extreme values of the input variable. However, there is more variation in the derivative values in the ranges $[-3, -1]$, and $[1, 3]$ than in a pure Gaussian density function. The mathematical representation of the ridgelet sieve network is given by the following system, with i^* input variables and k^* ridgelet sieves:

$$y_t^* = \sum_{i=1}^{i^*} \omega_i x_{i,t} \tag{2.41}$$

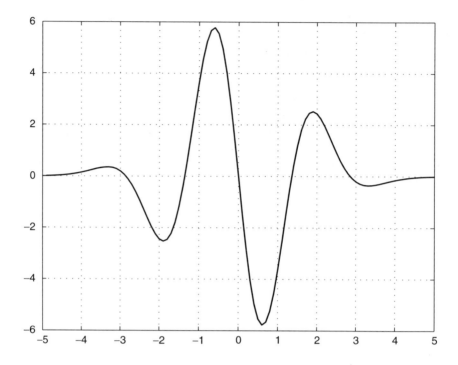

FIGURE 2.5. Ridgelet function

$$n_{k,t} = \alpha_k^{-1} \left(\beta_k \cdot y_t^* - \beta_{0,k} \right) \tag{2.42}$$

$$N_{k,t} = R(n_{k,t}) \tag{2.43}$$

$$y_t = \gamma_0 + \sum_{k=1}^{k^*} \frac{\gamma_k}{\sqrt{\alpha_k}} N_{k,t} \tag{2.44}$$

where α_k represents the scale while $\beta_{0,k}$ and β_k stand for the location and direction of the network, with $|\beta_l| = 1$.

2.4.5 Jump Connections

One alternative to the pure feedforward network or sieve network is a feedforward network with jump connections, in which the inputs x have direct linear links to output y, as well as to the output through the hidden layer of squashed functions. Figure 2.6 pictures a feedforward jump

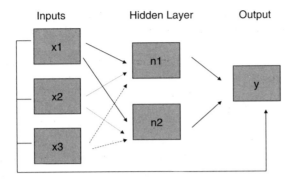

FIGURE 2.6. Feedforward neural network with jump connections

connection network with three inputs, one hidden layer, and two neurons $(i^* = 3, k^* = 2)$:

The mathematical representation of the feedforward network pictured in Figure 2.1, for logsigmoid activation functions, is given by the following system:

$$n_{k,t} = \omega_{k,0} + \sum_{i=1}^{i^*} \omega_{k,i} x_{i,t} \tag{2.45}$$

$$N_{k,t} = \frac{1}{1 + e^{-n_{k,t}}} \tag{2.46}$$

$$\hat{y}_t = \gamma_0 + \sum_{k=1}^{k^*} \gamma_k N_{k,t} + \sum_{i=1}^{i^*} \beta_i x_{i,t} \tag{2.47}$$

Note that the feedforward network with the jump connections increases the number of parameters in the network by j^*, the number of inputs. An appealing advantage of the feedforward network with jump connections is that it nests the pure linear model as well as the feedforward neural network. It allows the possibility that a nonlinear function may have a linear component as well as a nonlinear component. If the underlying relationship between the inputs and the output is a pure linear one, then only the direct jump connectors, given by the coefficient set $\{\beta_i\}$, $i = 1, \ldots, i^*$, should be significant. However, if the true relationship is a complex nonlinear one, then one would expect the coefficient sets $\{\omega\}$ and $\{\gamma\}$ to be highly significant, and the coefficient set $\{\beta\}$ to be relatively insignificant. Finally, the relationship between the input variables $\{x\}$ and the output variable

$\{y\}$ can be decomposed into linear and nonlinear components, and then we would expect all three sets of coefficients, $\{\beta\}, \{\omega\}$, and $\{\gamma\}$, to be significant.

A practical use of the jump connection network is as a useful test for neglected nonlinearities in a relationship between the input variables x and the output variable y. We take up this issue in the discussion of the Lee-White-Granger test. In this vein, we can also estimate a partitioned network. We first do linear least squares regression of the dependent variable y on the regressors, x, and obtain the residuals, e. We then set up a feedforward network in which the residuals from the linear regression become the dependent variable, while we use the same regressors as the input variables for the network. If there are indeed neglected nonlinearities in the linear regression, then the second-stage, partitioned network should have significant explanatory power.

Of course, the jump connection network and the partitioned linear and feedforward network should give equivalent results, at least in theory. However, as we discuss in the next section, due to problems of convergence to local rather than global optima, we may find that the results may be different, especially for networks with a large number of regressors and neurons in one or more hidden layers.

2.4.6 Multilayered Feedforward Networks

Increasing complexity may be approximated by making use of two or more hidden layers in a network architecture. Figure 2.7 pictures a feedforward network with two hidden layers, each having two neurons.

The representation of the network appearing in Figure 2.6 is given by the following system, with i^* input variables, k^* neurons in the first hidden

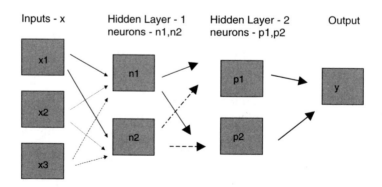

FIGURE 2.7. Feedforward network with two hidden layers

layer, and l^* neurons in the second hidden layer:

$$n_{k,t} = \omega_{k,0} + \sum_{i=1}^{i^*} \omega_{k,i} x_{i,t} \qquad (2.48)$$

$$N_{k,t} = \frac{1}{1 + e^{-n_{k,t}}} \qquad (2.49)$$

$$p_{l,t} = \rho_{l,0} + \sum_{k=1}^{k^*} \rho_{l,k} N_{k,t} \qquad (2.50)$$

$$P_{l,t} = \frac{1}{1 + e^{-p_{l,t}}} \qquad (2.51)$$

$$y_t = \gamma_0 + \sum_{l=1}^{l^*} \gamma_l P_{l,t} \qquad (2.52)$$

It should be clear that adding a second hidden layer increases the number of parameters to be estimated by the factor $(k^* + 1)(l^* - 1) + (l^* + 1)$, since the feedforward network with one hidden layer, with i^* inputs and k^* neurons, has $(i^* + 1)k^* + (k^* + 1)$ parameters, while a similar network with two hidden layers, with l^* neurons in the second hidden layer, has $(i^* + 1)k^* + (k^* + 1)l^* + (l^* + 1)$ hidden layers.

Feedforward networks with multiple hidden layers add complexity. They do so at the cost of more parameters to estimate, which use up valuable degrees of freedom if the sample size is limited, and at the cost of greater training time. With more parameters, there is also the likelihood that the parameter estimates may converge to a local, rather than global, optimum (we discuss this problem in greater detail in the next chapter). There has been a wide discussion about the usefulness of networks with more than one hidden layer. Dayhoff and DeLeo (2001), referring to earlier work by Hornik, Stinchcomb, and White (1989), make the following point on this issue:

A general function approximation theorem has been proven for three-layer neural networks. This result shows that artificial neural networks with two layers of trainable weights are capable of approximating any nonlinear function. This is a powerful computational property that is robust and has ramifications for many different applications of neural networks. Neural networks can approximate a multifactorial function in such a way that creating the functional form and fitting the function are performed at the same time, unlike nonlinear regression in which a fit is forced to a prechosen function. This capability gives neural networks a decided advantage over traditional statistical multivariate regression techniques.
[Dayhoff and DeLeo (2001), p. 1624].

In most situations, we can work with multilayer perceptron or jump-connection neural networks with one hidden layer and two or three neurons. We illustrate the advantage of a very simple neural network against a set of orthogonal polynomials in the next chapter.

2.4.7 Recurrent Networks

Another commonly used neural architecture is the Elman recurrent network. This network allows the neurons to depend not only on the input variables x, but also on their own lagged values. Thus the Elman network builds "memory" in the evolution of the neurons. This type of network is similar to the commonly used moving average (MA) process in time-series analysis. In the MA process, the dependent variable y is a function of observed inputs x as well as current and lagged values of an unobserved disturbance term or random shock, ϵ. Thus, a q-th order MA process has the following form:

$$y_t = \beta_0 + \sum_{i=1}^{i^*} \beta_i x_{i,t} + \epsilon_t + \sum_{j=1}^{q} \nu_j \widehat{\epsilon}_{t-j} \qquad (2.53)$$

$$\widehat{\epsilon}_{t-j} = y_{t-j} - \widehat{y}_{t-j} \qquad (2.54)$$

The q-dimensional coefficient set $\{\nu_j\}, j = 1, \ldots, q$, is estimated recursively. Estimation starts with ordinary least squares, eliminating the set of lagged disturbance terms, $\{\widehat{\epsilon}_{t-j}\}, j = 1, \ldots, q$. Then we take the set of residuals for the initial regression, $\{\widehat{\epsilon}\}$, as proxies for lagged $\{\widehat{\epsilon}_{t-j}\}, j = 1, \ldots, q$, and estimate the parameters $\{\beta_i\}, i = 0, \ldots, i^*$, as well as the set of coefficients of the lagged disturbances, $\{\nu_j\}, j = 1, \ldots, q$. The process continues over several steps until convergence is achieved and when further iterations produce little or no change in the estimated coefficients.

In a similar fashion, the Elman network makes use of lagged as well as current values of unobserved unsquashed neurons in the hidden layer. One such Elman recurrent network appears in Figure 2.8, with three inputs, two neurons in one hidden layer, and one output. In the estimation of both Elman networks and MA processes, it is necessary to use a multistep estimation procedure. We start with initializing the vector of lagged neurons with lagged neuron proxies from a simple feedforward network. Then we estimate their coefficients and recalculate the vector of lagged neurons. Parameter values are re-estimated in a recursive fashion. The process continues until convergence takes place.

Note that the inputs, neurons, and output boxes have time labels for the current period, t, or the lagged period, $t - 1$. The Elman network is thus a network specific to data that have a time dimension. The feedforward

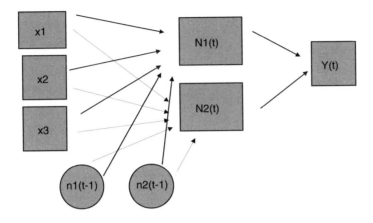

FIGURE 2.8. Elman recurrent network

network, on the other hand, may be used for cross-section data, which are not dimensioned by time, as well as time-series data.

The following system represents the recurrent Elman network illustrated in Figure 2.8:

$$n_{k,t} = \omega_{k,0} + = \omega_{k,0} + \sum_{i=1}^{i^*} \omega_{k,i} x_{i,t} + \sum_{k=1}^{k^*} \phi_k n_{k,t-1} \qquad (2.55)$$

$$N_{k,t} = \frac{1}{1 + e^{-n_{i,t}}} \qquad (2.56)$$

$$y_t = \gamma_0 + \sum_{k=1}^{k^*} \gamma_k N_{k,t}$$

Note that the recurrent Elman network is one in which the lagged hidden-layer neurons feed back into the current hidden layer of neurons. However, the lagged neurons do so before the logsigmoid activation function is applied to them — they enter as lags in their unsquashed state. The recurrent network thus has an indirect feedback effect from the lagged unsquashed neurons to the current neurons, not a direct feedback from lagged neurons to the level of output. The moving-average time-series model, on the other hand, has a direct feedback effect, from lagged disturbance terms to the level of output y_t. Despite the recursive estimation process for obtaining proxies of nonobserved data, the recurrent network differs in this one important respect from the moving-average time-series model.

The Elman network is a way of capturing memory in financial markets, particularly for forecasting high-frequency data such as daily, intra-daily, or even real-time returns in foreign exchange or share markets. While the use of lags certainly is one way to capture memory, memory may also show up in the way the nonlinear structure changes through time. The use of the Elman network, in which the lagged neurons feed back into the current neurons, is a very handy way to model this type of memory structure, in which the hidden layer itself changes through time, due to feedback from past neurons.

The Elman network is an explicit dynamic network. The feedforward network is usually regarded as a static network, in which a given set of input variables at time t are used to forecast a target output variable at time t. Of course, the input variables used in the feedforward network may be lagged values of the output variable, so that the feedforward network becomes dynamic by redefinition of the input variables. The Elman network, by contrast, allows another dynamic structure beyond incorporating lagged dependent or output variables, $y_{t-1}, \ldots, y_{t-k}$, as current input variables. Moreover, as Mandic and Chambers (2001) point out, restricting memory or dynamic structure in the feedforward network only to the input structure may lead to an unnecessarily large number of parameters. While recurrent networks may be functionally equivalent to feedforward networks with only lagged input variables, they generally have far fewer parameters, which, of course, speeds up the estimation or training process.

2.4.8 Networks with Multiple Outputs

Of course, a feedforward network (or Elman network) can have multiple outputs. Figure 2.9 shows one such feedforward network architecture, with three inputs, two neurons, and two outputs. The representation of the feedforward network architecture is given by the following system:

$$n_{k,t} = \omega_{k,0} + \sum_{i=1}^{i^*} \omega_{k,i} x_{i,t} \tag{2.57}$$

$$N_{k,t} = \frac{1}{1 + e^{-n_{k,t}}} \tag{2.58}$$

$$y_{1,t} = \gamma_{1,0} + \sum_{k=1}^{k^*} \gamma_{1,k} N_{k,t} \tag{2.59}$$

$$y_{2,t} = \gamma_{2,0} + \sum_{k=1}^{k^*} \gamma_{2,k} N_{k,t} \tag{2.60}$$

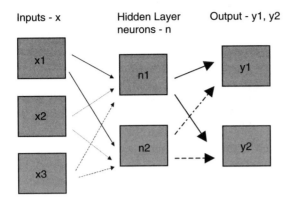

FIGURE 2.9. Feedforward network multiple outputs

We see in this system that the addition of one additional output in the feedforward network requires additional $(k^* + 1)$ parameters, equal to the number of neurons on the hidden layer plus an additional constant term. Thus, adding more output variables to be predicted by the network requires additional parameters which depend on the number of neurons in the hidden layer, not on the number of input variables.

By contrast, a linear model depending on k regressors or arguments plus a constant would require additional $k + 1$ parameters — essentially a new separate regression — for each additional output variable. Similarly, a polynomial approximation would require a doubling of the number of parameters for each additional output.

The use of a single feedforward network with multiple outputs makes sense, of course, when the outputs of the network are closely related or dependent on the same set of input variables. This type of network is especially useful, as well as economical or parsimonious in terms of parameters, when we are forecasting a specific variable, such as inflation, at different horizons. The set of input variables would be the usual determinants of inflation, such as lags of inflation, and demand and cost variables. The output variables could be inflation forecasts at one-month, quarterly, six-month, and one-year horizons.

Another application would be a forecast of the term structure of interest rates. The output variables would be forecasts of interest rates for maturities of three months, six months, one year, and perhaps two years, while the input variables would be the usual determinants of interest rates, such as monetary growth rates, lagged inflation rates, and foreign interest rates.

Finally, classification networks, discussed below, are a very practical application of multiple-output networks. In this type of model, for example, we may wish to classify outcomes as a probability of low, medium, or high risk. We would have two outputs for the probability of low and medium risk, and the high-risk case would simply be one minus the two probabilities.

2.5 Neural Network Smooth-Transition Regime Switching Models

While the networks discussed above are commonly used approximators, an important question remains: How can we adapt these networks for addressing important and recurring issues in empirical macroeconomics and finance? In particular, researchers have long been concerned with structural breaks in the underlying data-generating process for key macroeconomic variables such as GDP growth or inflation. Does one regime or structure hold when inflation is high and another when inflation is low or even below zero? Similarly, do changes in GDP have one process in recession and another in recovery? These are very important questions for forecasting and policy analysis, since they also involve determining the likelihood of breaking out of a deflation or recession regime.

There have been many macroeconomic time-series studies based on *regime switching models*. In these models, one set of parameters governs the evolution of the dependent variable, for example, when the economy is in recovery or positive growth, and another set of parameters governs the dependent variable when the economy is in recession or negative growth. The initial models incorporated two different linear regimes, switching between periods of recession and recovery, with a discrete Markov process as the transition function from one regime to another [see Hamilton (1989, 1990)]. Similarly, there have been many studies examining non-linearities in business cycles, which focus on the well-observed asymmetric adjustments in times of recession and recovery [see Teräsvirta and Anderson (1992)]. More recently, we have seen the development of smooth-transition regime switching models, discussed in Frances and van Dijk (2000), originally developed by Teräsvirta (1994), and more generally discussed in van Dijk, Teräsvirta, and Franses (2000).

2.5.1 Smooth-Transition Regime Switching Models

The smooth-transition regime switching framework for two regimes has the following form:

$$y_t = \alpha_1 \mathbf{x}_t \cdot \Psi(y_{t-1}; \theta, c) + \alpha_2 \mathbf{x}_t \cdot [1 - \Psi(y_{t-1}; \theta, c)] \qquad (2.61)$$

where $\mathbf{x}_t$ is the set of regressors at time t, α_1 represents the parameters in state 1, and α_2 is the parameter vector in state 2. The transition function Ψ, which determines the influence of each regime or state, depends on the value of y_{t-1} as well as a smoothness parameter vector θ and a threshold parameter c. Franses and van Dijk (2000, p. 72) use a logistic or logsigmoid specification for $\Psi(y_{t-1}; \theta, c)$:

$$\Psi(y_{t-1}; \theta, c) = \frac{1}{1 + \exp[-\theta(y_{t-1} - c)]} \tag{2.62}$$

Of course, we can also use a cumulative Gaussian function instead of the logistic function. Measures of Ψ are highly useful, since they indicate the likelihood of continuing in a given state. This model, of course, can be extended to multiple states or regimes [see Franses and van Dijk (2000), p. 81].

2.5.2 Neural Network Extensions

One way to model a smooth-transition regime switching framework with neural networks is to adapt the feedforward network with jump connections. In addition to the direct linear links from the inputs or regressors x to the dependent variable y, holding in all states, we can model the regime switching as a jump-connection neural network with one hidden layer and two neurons, one for each regime. These two regimes are weighted by a logistic connector which determines the relative influence of each regime or neuron in the hidden layer. This system appears in the following equations:

$$y_t = \alpha \mathbf{x}_t + \beta \{ [\Psi(y_{t-1}; \theta, c)] G(\mathbf{x}_t; \kappa) +$$

$$[1 - \Psi(y_{t-1}; \theta, c)] H(\mathbf{x}_t; \lambda) \} + \eta_t \tag{2.63}$$

where $\mathbf{x}_t$ is the vector of independent variables at time t, and α represents the set of coefficients for the direct link. The functions $G(\mathbf{x}_t; \kappa)$ and $H(\mathbf{x}_t; \lambda)$, which capture the two regimes, are logsigmoid and have the following representations:

$$G(\mathbf{x}_t; \kappa) = \frac{1}{1 + \exp[-\kappa \mathbf{x}_t]} \tag{2.64}$$

$$H(\mathbf{x}_t; \lambda) = \frac{1}{1 + \exp[-\lambda \mathbf{x}_t]} \tag{2.65}$$

where the coefficient vectors κ and λ are the coefficients for the vector $\mathbf{x}_t$ in the two regimes, $G(\mathbf{x}_t; \kappa)$ and $H(\mathbf{x}_t; \lambda)$.

Transition function Ψ, which determines the influence of each regime, depends on the value of y_{t-1} as well as the parameter vector θ and a threshold parameter c. As Franses and van Dyck (2000) point out, the parameter θ determines the smoothness of the change in the value of this function, and thus the transition from one regime to another regime.

This neural network regime switching system encompasses the linear smooth-transition regime switching system. If nonlinearities are not significant, then the parameter β will be close to zero. The linear component may represent a core process which is supplemented by nonlinear regime switching processes. Of course there may be more regimes than two, and this system, like its counterpart above, may be extended to incorporate three or more regimes. However, for most macroeconomic and financial studies, we usually consider two regimes, such as recession and recovery in business cycle models or inflation and deflation in models of price adjustment.

As in the case of linear regime switching models, the most important payoff of this type of modeling is that we can forecast more accurately not only the dependent variable, but also the probability of continuing in the same regime. If the economy is in deflation or recession, given by the $H(\mathbf{x}_t; \lambda)$ neuron, we can determine if the likelihood of continuing in this state, $1 - \Psi(y_{t-1}; \theta, c)$, is close to zero or one, and whether this likelihood is increasing or decreasing over time.[9]

Figure 2.10 displays the architecture of this network for three input variables.

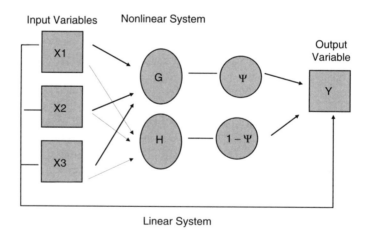

FIGURE 2.10. NNRS model

---

[9]In succeeding chapters, we compare the performance of the neural network smooth-transition regime switching system with that of the linear smooth-transition regime switching model and the pure linear model.

2.6 Nonlinear Principal Components: Intrinsic Dimensionality

Besides forecasting specific target or output variables, which are determined or predicted by specific input variables or regressors, we may wish to use a neural network for dimensionality reduction or for distilling a large number of potential input variables into a smaller subset of variables that explain most of the variation in the larger data set. Estimation of such networks is called *unsupervised training*, in the sense that the network is not evaluated or supervised by how well it predicts a specific readily observed target variable.

Why is this useful? Many times, investors make decisions on the basis of a signal from the market. In point of fact, there are many markets and many prices in financial markets. Well-known indicators such as the Dow-Jones Industrial Average, the Standard and Poor 500, or the National Association of Security Dealers' Automatic Quotations (NASDAQ) are just that, indices or averages of prices of specific shares or all the shares listed on the exchanges. The problem with using an index based on an average or weighted average is that the market may not be clustered around the average.

Let's take a simple example: grades in two classes. In one class, half of the students score 80 and the other half score 100. In another class, all of the students score 90. Using only averages as measures of student performances, both classes are identical. Yet in the first class, half of the students are outstanding (with a grade of 100) and the other half are average (with a grade of 80). In the second class, all are above average, with a grade of 90. We thus see the problem of measuring the *intrinsic dimensionality* of a given sample. The first class clearly needs two measures to explain satisfactorily the performance of the students, while one measure is sufficient for the second class.

When we look at the performance of financial markets as a whole, just as in the example of the two classes, we note that single indices can be very misleading about what is going on. In particular, the market average may appear to be stagnant, but there may be some very good performers which the overall average fails to signal.

In statistical estimation and forecasting, we often need to reduce the number of regressors to a more manageable subset if we wish to have a sufficient number of degrees of freedom for any meaningful inference. We often have many candidate variables for indicators of real economic activity, for example, in studies of inflation [see Stock and Watson (1999)]. If we use all of the possible candidate variables as regressors in one model, we bump up against the "curse of dimensionality," first noted by Bellman (1961). This "curse" simply means that the sample size needed to estimate a model with a given degree of accuracy grows exponentially with the number of variables in the model.

Another reason for turning to dimensionality reduction schemes, especially when we work with high-frequency data sets, is the empty space phenomenon. For many periods, if we use very small time intervals, many of the observations for the variables will be at zero values. Such a set of variables is called a *sparse data set*. With such a data set estimation becomes much more difficult, and dimensionality reduction methods are needed.

2.6.1 Linear Principal Components

The linear approach to reducing a larger set of variables into a smaller subset of signals from a large set of variables is called *principal components analysis* (PCA). PCA identifies linear projections or combinations of data that explain most of the variation of the original data, or extract most of the information from the larger set of variables, in decreasing order of importance. Obviously, and trivially, for a data set of K vectors, K linear combinations will explain the total variation of the data. But it may be the case that only two or three linear combinations or principal components may explain a very large proportion of the variation of the total data set, and thus extract most of the useful information for making decisions based on information from markets with large numbers of prices.

As Fotheringhame and Baddeley (1997) point out, if the underlying true structure interrelating the data is linear, then a few principal components or linear combinations of the data can capture the data "in the most succinct way," and the resulting components are both uncorrelated and independent [Fotheringhame and Baddeley (1997), p. 1].

Figure 2.11 illustrates the structure of principal components mapping. In this figure, four input variables, $x1$ through $x4$, are mapped into identical output variables $x1$ through $x4$, by H units in a single hidden layer. The H units in the hidden layer are linear combinations of the input variables. The output variables are themselves linear combinations of the H units. We can call the mapping from the inputs to the H-units a "dimensionality reduction mapping," while the mapping from the H-units to the output variables is a "reconstruction mapping."[10]

The method by which the coefficients linking the input variables to the H units are estimated is known as orthogonal regression. Letting $X = [x_1, \ldots, x_k]$ be a dimension T by k matrix of variables we obtain the following eigenvalues λ_x and eigenvectors ν_x through the process of orthogonal

---

[10]See Carreira-Perpinan (2001) for further discussion of dimensionality reduction in the context of linear and nonlinear methods.

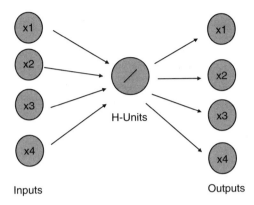

FIGURE 2.11. Linear principal components

regression through calculation of eigenvalues and eigenvectors:

$$[X'X - \lambda_x I]\nu_x = 0 \qquad (2.66)$$

For a set of k regressors, there are, of course, at most k eigenvalues and k eigenvectors. The eigenvalues are ranked from the largest to the smallest. We use the eigenvector ν_x associated with the largest eigenvalue to obtain the first principal component of the matrix X. This first principle component is simply a vector of length T, computed as a weighted average of the k-columns of X, with the weighting coefficients being the elements of ν_x. In a similar manner, we may find second and third principal components of the input matrix by finding the eigenvector associated with the second and third largest eigenvalues of the matrix X, and multiplying the matrix by the coefficients from the associated eigenvectors.

The following system of equations shows how we calculate the principle components from the ordered eigenvalues and eigenvectors of a T-by-k dimension matrix X:

$$\left(X'X - \begin{bmatrix} \lambda_x^1 & 0 & 0\ldots 0 \\ 0 & \lambda_x^2 & 0\ldots 0 \\ 0 & 0 & 0\ldots \lambda_x^k \end{bmatrix} \cdot I_k \right) [\nu_x^1 \ \nu_x^2 \ldots \nu_x^k] = 0$$

The total explanatory power of the first two or three sets of principal components for the entire data set is simply the sum of the two or three largest eigenvalues divided by the sum of all of eigenvalues.

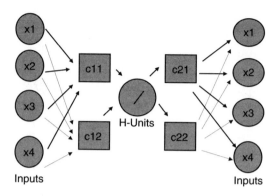

FIGURE 2.12. Neural principal components

2.6.2 Nonlinear Principal Components

The neural network structure for nonlinear principal components analysis (NLPCA) appears in Figure 2.12, based on the representation in Fotheringhame and Baddeley (1997).

The four input variables in this network are encoded by two intermediate logsigmoid units, $C11$ and $C12$, in a dimensionality reduction mapping. These two encoding units are combined linearly to form H neural principal components. The H-units in turn are decoded by two decoding logsigmoid units $C21$ and $C22$, in a reconstruction mapping, which are combined linearly to regenerate the inputs as the output layers.[11] Such a neural network is known as an *auto-associative mapping*, because it maps the input variables $x_1, \ldots, x_4$ into themselves.

Note that there are two logsigmoidal unities, one for the dimensionality reduction mapping and one for the reconstruction mapping.

Such a system has the following representation, with **EN** as an encoding neuron and **DN** as a decoding neuron. Letting X be a matrix with K columns, we have J encoding and decoding neurons, and P nonlinear principal components:

$$EN_j = \sum_{k=1}^{K} \alpha_{j,k} X_k$$

$$\mathbf{EN}_j = \frac{1}{1 + \exp(-EN_j)}$$

---

[11]Fotheringhame and Baddeley (1997) point out that although it is not strictly required, networks usually have equal numbers in the encoding and decoding layers.

$$H_p = \sum_{j=1}^{J} \beta_{p,j} \mathbf{EN}_j$$

$$DN_j = \sum_{p=1}^{P} \gamma_{j,p} H_p$$

$$\mathbf{DN}_j = \frac{1}{1 + \exp(-DN_j)}$$

$$\widehat{X}_k = \sum_{j=1}^{J} \delta_{k,j} \mathbf{DN}_j$$

The coefficients of the network link the input variables x to the encoding neurons $C11$ and $C12$, and to the nonlinear principal components. The parameters also link the nonlinear principal components to the decoding neurons $C21$ and $C22$, and the decoding neurons to the same input variables x. The natural way to start is to take the sum of squared errors for each of the predicted values of x, denoted by $\widehat{x}$ and the actual values. The sum of the total squared errors for all of the different x's is the object of minimization, as shown in Equation 2.67:

$$Min \sum_{j=1}^{k} \sum_{t=1}^{T} [x_{jt} - \widehat{x}_{jt}]^2 \tag{2.67}$$

where k is the number of input variables and T is the number of observations. This procedure in effect gives an equal weight to all of the input categories of x. However, some of the inputs may be more volatile than others, and thus harder to accurately predict as than others. In this case, it may not be efficient to give equal weight to all of the variables, since the computer will be working equally hard to predict inherently less predictable variables as it is for more predictable variables. We would like the computer to spend more time where there is a greater chance of success. In robust regression, we can weight the different squared errors of the input variables differently, giving less weight to those inputs that are inherently more volatile or less predictable and more weight to those that are less volatile and thus easier to predict:

$$Min[v\widehat{\Sigma}^{-1}v'] \tag{2.68}$$

where α_j is the weight given to each of the input variables. This weight is determined during the estimation process itself. As each of the errors is computed for the different input variables, we form the matrix $\widehat{\Sigma}$ during the estimation process:

$$
E = \begin{bmatrix}
\widehat{e}_{11}\widehat{e}_{21} \ldots \widehat{e}_{k1} \\
\widehat{e}_{12}\widehat{e}_{22} \ldots \widehat{e}_{k2} \\
\vdots \quad \ddots \\
\widehat{e}_{1T}\widehat{e}_{2T} \ldots \widehat{e}_{kT}
\end{bmatrix}
\tag{2.69}
$$

$$
\widehat{\Sigma} = E'E
\tag{2.70}
$$

where $\widehat{\Sigma}$ is the variance–covariance matrix of the residuals and v is the row vector of the sum of squared errors:

$$
v_t = [\widehat{e}_{1t}\widehat{e}_{2t} \ldots \widehat{e}_{kt}]
\tag{2.71}
$$

This type of robust estimation, of course, is applicable to any model having multiple target or output variables, but it is particularly useful for nonlinear principal components or auto-associative maps, since valuable estimation time will very likely be wasted if equal weighting is given to all of the variables. Of course, each $\widehat{e}_{kt}$ will change during the course of the estimation process or training iterations. Thus $\widehat{\Sigma}$ will also change and initially not reflect the true or final covariance weighting matrix. Thus, for the initial stages of the training, we set $\widehat{\Sigma}$ equal to the identity matrix of dimension k, I_k. Once the nonlinear network is trained, the output is the space spanned by the first H nonlinear principal components.

Estimation of a nonlinear dimensionality reduction method is much slower than that of linear principal components. We show, however, that this approach is much more accurate than the linear method when we have to make decisions in real time. In this case, we do not have time to update the parameters of the network for reducing the dimension of a sample. When we have to rely on the parameters of the network from the last period, we show that the nonlinear approach outperforms the linear principal components.

2.6.3   Application to Asset Pricing

The H principal component units from linear orthogonal regression or neural network estimation are particularly useful for evaluating expected or required returns for new investment opportunities, based on the capital asset pricing model, better known as the CAPM. In its simplest form, this theory requires that the minimum required return for any asset or portfolio k, $\widetilde{r}_k$, net of the risk-free rate r_f, is proportional, by a factor β_k, to the difference between the observed market return, r_m, less the risk-free rate:

$$\tilde{r}_k = r_f + \beta_k[r_m - r_f] \tag{2.72}$$

$$\beta_k = \frac{Cov(r_k, r_m)}{Var(r_m)} \tag{2.73}$$

$$r_{k,t} = \tilde{r}_{k,t} + \epsilon_t \tag{2.74}$$

The coefficient β_k is widely known as the CAPM beta for an asset or portfolio return k, and is computed as the ratio of the covariance of the returns on asset k with the market return, divided by the variance of the return on the market. This beta, of course, is simply a regression coefficient, in which the return on asset k, r_k, less the risk-free rate, r_f, is regressed on the market rate, r_m, less the same risk-free rate. The observed market return at time t, $r_{k,t}$, is assumed to be the sum of two components: the required return, $\tilde{r}_{k,t}$, and an unexpected noise or random shock, ϵ_t. In this CAPM literature, the actual return on any asset $r_{k,t}$ is a compensation for risk. The required return $\tilde{r}_{k,t}$ represents diversifiable risk in financial markets, while the noise term represents nondiversifiable idiosyncratic risk at time t.

The appeal of the CAPM is its simplicity in deriving the minimum expected or required return for an asset or investment opportunity. In theory, all we need is information about the return of a particular asset k, the market return, the risk-free rate, and the variance and covariance of the two return series. As a decision rule, it is simple and straightforward: if the current observed return on asset k at time t, $r_{k,t}$, is greater than the required return, $\tilde{r}_k$, then we should invest in this asset.

However, the limitation of the CAPM is that it identifies the market return with only one particular market return. Usually the market return is an index, such as the Standard and Poor or the Dow-Jones, but for many potential investment opportunities, these indices do not reflect the relevant or benchmark market return. The market average is not a useful signal representing the news and risks coming from the market. Not surprisingly, the CAPM model does not do very well in explaining or predicting the movement of most asset returns.

The arbitrage pricing theory (APT) was introduced by Ross (1976) as an alternative to the CAPM. As Campbell, Lo, and MacKinlay (1997) point out, the APT provides an approximate relation for expected or required asset returns by replacing the single benchmark market return with a number of unidentified factors, or principal components, distilled from a wide set of asset returns observed in the market.

The intertemporal capital asset pricing model (ICAPM) developed by Merton (1973) differs from the APT in that it specifies the benchmark market return index as one argument determining the required return, but allows additional arguments or state variables, such as the principal components distilled from a wider set of returns. These arise, as Campbell, Lo, and MacKinlay (1997) point out, from investors' demand to hedge uncertainty about further investment opportunities.

In practical terms, as Campbell, Lo, and MacKinlay also note, it is not necessary to differentiate the APT from the ICAPM. We may use one observed market return as one variable for determining the required return. But one may include other arguments as well, such as macroeconomic indicators that capture the systematic risk of the economy. The final remaining arguments can be the principal components, either from the linear or neural estimation, distilled from a wide set of observed asset returns.

Thus, the required return on asset k, $\widetilde{r}_k$, can come from a regression of these returns, on one overall market index rate of return, on a set of macroeconomic variables (such as the yield spread between long- and short-term rates for government bonds, the expected and unexpected inflation rates, industrial production growth, and the yield between corporate high and low-grade bonds) and on a reasonably small set of principal components obtained from a wide set of returns observed in the market. Campbell, Lo, and MacKinlay cite research that suggests that five would be an adequate number of principal components to compute from the overall set of returns observed in the market.

We can of course combine the forecasts of the CAPM, the APT, and the nonlinear autoassociative maps associated with the nonlinear principal component forecasts with a thick model. Granger and Jeon (2001) describe *thick modeling* as "using many alternative specifications of similar quality, using each to produce the output required for the purpose of the modeling exercise," and then combining or synthesizing the results [Granger and Jeon (2001), 3].

Finally, as we discuss later, a very useful application — likely the most useful application — of nonlinear principal components is to distill information about the underlying volatility dynamics from observed data on implied volatilities in markets for financial derivatives. In particular, we can obtain the implied volatility measures on all sorts of options, and swap-options or "swaptions" of maturities of different lengths, on a daily basis. What is important for market participants to gauge is the behavior of the market as a whole: From these diverse signals, volatilities of different maturities, is the riskiness of the market going up or down? We show that for a variety of implied volatility data, one nonlinear principal component can explain a good deal of the overall market riskiness, where it takes two or more linear principal components to achieve the same degree of explanatory power. Needless to say, one measure for summing up market developments is much better than two or more.

While the CAPM, APT, and ICAPM are used for making decisions about required returns, nonlinear principal components may also be used in a dynamic context, in which lagged variables may include lagged linear or nonlinear principal components for predicting future rates of return for any asset. Similarly, the linear or nonlinear principal component may be used to reduce a larger number of regressors to a smaller, more manageable number of regressors for any type of model. A pertinent example would be to distill a set of principal components from a wide set of candidate variables that serve as leading indicators for economic activity. Similarly, linear or nonlinear principal components distilled from the wider set of leading indicators may serve as the proxy variables for overall aggregate demand in models of inflation.

2.7 Neural Networks and Discrete Choice

The analysis so far assumes that the dependent variable, y, to be predicted by the neural network, is a continuous random variable rather than a discrete variable. However, there are many cases in financial decision making when the dependent variable is discrete. Examples are easy to find, such as classifying potential loans as low and acceptable risk or high and unacceptable. Another is the likelihood that a particular credit card transaction is a true or a fraudulent charge.

The goal of this type of analysis is to classify data, as accurately as possible, into membership in two groups, coded as 0 or 1, based on observed characteristics. Thus, information on current income, years in current job, years of ownership of a house, and years of education, may help classify a particular customer as an acceptable or high-risk case for a new car loan. Similarly, information about the time of day, location, and amount of a credit card charge, as well as the normal charges of a particular card user, may help a bank security officer determine if incoming charges are more likely to be true and classified as 0, or fraudulent and classified as 1.

2.7.1 Discriminant Analysis

The classical linear approach for classification based on observed characteristics is linear discriminant analysis. This approach takes a set of k-dimensional characteristics from observed data falling into two groups, for example, a group that paid its loans on schedule and another that became arrears in loan payments. We first define the matrices X_1, X_2, where the rows of each X_i represent a series of k-different characteristics of the members of each group, such as a low-risk or a high-risk group. The relevant characteristics may be age, income, marital status, and years in current employment. Discriminant analysis proceeds in three steps:

1. Calculate the means of the two groups, $\overline{X}_1, \overline{X}_2$, as well as the variance–covariance matrices, $\widehat{\Sigma}_1, \widehat{\Sigma}_2$.

2. Compute the pooled variance, $\widehat{\Sigma} = \left(\frac{n_1-1}{n_1+n_2-2}\right)\widehat{\Sigma}_1 + \left(\frac{n_2-1}{n_1+n_2-2}\right)\widehat{\Sigma}_2$, where n_1, n_2 represent the population sizes in groups 1 and 2.

3. Estimate the coefficient vector, $\widehat{\beta} = \widehat{\Sigma}^{-1}\left[\overline{X}_1 - \overline{X}_2\right]$.

4. With the vector $\widehat{\beta}$, examine the characteristics of a new set of characteristics for classification in either the low-risk or high-risk sets, X_1 or X_2. Defining the net set of characteristics, x_i, we calculate the value: $\widehat{\beta}x_i$. If this value is closer to $\widehat{\beta}\overline{X}_1$ than to $\widehat{\beta}\overline{X}_2$, then we classify x_i as belonging to the low-risk group X_1. Otherwise, it is classified as being a member of X_2.

Discriminant analysis has the advantage of being quick, and has been widely used for an array of interesting financial applications.[12] However, it is a simple linear method, and does not take into account any assumptions about the distribution of the dependent variable used in the classification. It classifies a set of characteristics $\widetilde{X}$ as belonging to group 1 or 2 simply by a distance measure. For this reason it has been replaced by the more commonly used logistic regression.

2.7.2 Logit Regression

Logit analysis assumes the following relation between probability p_i of the binary dependent variable y_i, taking values zero or one, and the set of k explanatory variables x:

$$p_i = \frac{1}{1 + e^{-[x_i\beta + \beta_0]}} \tag{2.75}$$

To estimate the parameters β and β_0, we simply maximize the following log-likelihood function Λ with respect to the parameter vector β:

$$\underset{<\beta>}{Max}\Lambda = \prod (p_i)^{y_i}(1-p_i)^{1-y_i} \tag{2.76}$$

$$= \prod \left(\frac{1}{1 + e^{-[x_i\beta+\beta_0]}}\right)^{y_i}\left(\frac{e^{-[x_i\beta+\beta_0]}}{1 + e^{-[x_i\beta+\beta_0]}}\right)^{1-y_i} \tag{2.77}$$

where y_i represents the observed discrete outcomes.

---

[12]For example, see Altman (1981).

For optimization, it is sometimes easier to optimize the log-likelihood function $\ln(\Lambda)$:

$$\underset{<\beta>}{Max} \ln(\Lambda) = y_i \ln(p_i) + (1 - y_i) \ln(1 - p_i) \tag{2.78}$$

The k dimensional coefficient vector β does not represent a set of partial derivatives with respect to characteristics x_k. The partial derivative comes from the following expression:

$$\frac{\partial p_i}{\partial x_{i,k}} = \frac{e^{x_i\beta + \beta_0}}{(1 + e^{x_i\beta + \beta_0})^2} \beta_k \tag{2.79}$$

The partial derivatives are of particular interest if we wish to identify critical characteristics that increase or decrease the likelihood of being in a particular state or category, such as representing a risk of default on a loan.[13,14]

The usual way to evaluate this logistic model is to examine the percentage of correct predictions, both true and false, set at 1 and 0, on the basis of the expected value. Setting the estimated p_i at 0 or 1 depends on the choice of an appropriate threshold value. If the estimated probability or expected value p_i is greater than .5, then p_i is rounded to 1, and expected to take place. Otherwise, it is not expected to occur.[15]

2.7.3 Probit Regression

Probit models are also used: these models simply use the cumulative Gaussian normal distribution rather than the logistic function for calculating the probability of being in one category or not:

$$p_i = \Phi(x_i\beta + \beta_0)$$

$$= \int_{-\infty}^{x_i\beta + \beta_0} \phi(t)dt$$

where the symbol Φ is simply the cumulative standard distribution, while the lower case symbol, ϕ, as before, represents the standard normal density function. We maximize the same log-likelihood function. The partial

---

[13]In many cases, a risk-averse decision maker may take a more conservative approach. For example, if the risk of having serious cancer exceeds .3, the physician may wish to diagnose the patient as a "high risk," warranting further diagnosis.

[14]More discussion appears in Section 2.7.4 about the computation of partial derivatives in nonlinear neural network regression.

[15]Further discussion appears in Section 2.8 about evaluating the success of a nonlinear regression.

derivatives, however, come from the following expression:

$$\frac{\partial p_i}{\partial x_{i,k}} = \phi(x_i\beta + \beta_0)\beta_k \tag{2.80}$$

Greene (2000) points out that the logistic distribution is similar to the normal one, except in the tails. However, he points out that it is difficult to justify the choice of one distribution or another on "theoretical grounds," and for most cases, "it seems not to make much difference" [Greene (2000), p. 815].

2.7.4   Weibull Regression

The Weibull distribution is an asymmetric distribution, strongly negatively skewed, approaching zero only slowly, and 1 more rapidly than the probit and logit models:

$$p_i = 1 - \exp(-\exp(x_i\beta + \beta_0)) \tag{2.81}$$

This distribution is used for classification in survival analysis and comes from "extreme value theory." The partial derivative is given by the following equation:

$$\frac{\partial p_i}{\partial x_{i,k}} = \exp(x_i\beta + \beta_0)\exp(-(x_i\beta + \beta_0))\beta_k \tag{2.82}$$

This distribution is also called the Gompertz distribution and the regression model is called the Gompit model.

2.7.5   Neural Network Models for Discrete Choice

Logistic regression is a special case of neural network regression for binary choice, since the logistic regression represents a neural network with one hidden neuron. The following adapted form of the feedforward network may be used for a discrete binary choice model, predicting probability p_i for a network with k^* input characteristics and j^* neurons:

$$n_{j,i} = \omega_{j,0} + \sum_{k=1}^{k^*} \omega_{j,k} x_{k,i} \tag{2.83}$$

$$N_{j,i} = \frac{1}{1 + e^{-n_{j,i}}} \tag{2.84}$$

$$\widetilde{p}_i = \sum_{j=1}^{j^*} \gamma_j N_{j,i} \tag{2.85}$$

$$\sum_{j=1}^{j^*} \gamma_j = 1, \gamma_j \geq 0$$

Note that the probability $\widetilde{p}_i$ is a weighted average of the logsigmoid neurons $N_{j,i}$, which are bounded between 0 and 1. Since the final probability is also bounded in this way, the final probability is a weighted average of these neurons. As in logistic regression, the coefficients are obtained by maximizing the product of likelihood function, given the preceding (or the sum of the log-likelihood function).

The partial derivatives of the neural network discrete choice models are given by the following expression:

$$\frac{\partial p_i}{\partial x_{i,k}} = \sum_{j=1}^{j^*} \gamma_j N_{j,i}(1 - N_{j,i})\omega_{j,k}$$

2.7.6   Models with Multinomial Ordered Choice

It is straightforward to extend the logit and neural network models to the case of multiple discrete choices or classification into three or more outcomes. In this case, logit regression is known as *logistic estimation*. For example, a credit officer may wish to classify potential customers into safe, low-risk, and high-risk categories based on a net of characteristics, x_k.

One direct approach for such a classification is a nested classification. One can use the logistic or neural network model to separate the normal categories from the absolute default or high-risk categories, with a first-stage estimation. Then, with the remaining normal data, one can separate the categories into low-risk and higher-risk categories.

However, there are many cases in financial decision making where there are multiple categories. Bond ratings, for example, are often in three or four categories. Thus, one might wish to use logistic or neural network classification to predict which type of category a particular firm's bond may fall into, given the characteristics of the particular firm, from observable market data and current market classifications or bond ratings.

In this case, using the example of three outcomes, we use the softmax function to compute p_1, p_2, p_3 for each observation i:

$$P_{1,i} = \frac{1}{1 + e^{-[x_i\beta_1 + \beta_{10}]}} \tag{2.86}$$

$$P_{2,i} = \frac{1}{1 + e^{-[x_i\beta_2 + \beta_{20}]}} \tag{2.87}$$

$$P_{3,i} = \frac{1}{1 + e^{-[x_i\beta_3 + \beta_{30}]}} \tag{2.88}$$

The probabilities of falling in category 1, 2, or 3 come from the cumulative probabilities:

$$p_{1,i} = \frac{P_{1,i}}{\sum_{j=1}^{3} P_{j,i}} \tag{2.89}$$

$$p_{2,i} = \frac{P_{2,i}}{\sum_{j=1}^{3} P_{j,i}} \tag{2.90}$$

$$p_{3,i} = \frac{P_3}{\sum_{j=1}^{3} P_{j,i}} \tag{2.91}$$

Neural network models yield the cumulative probabilities in a similar manner. In this case there are m^* neurons in the hidden layer, k^* inputs, and j probability outputs at each observation i, for i^* observations:

$$n_{m,i} = \omega_{m,0} + \sum_{k=1}^{k^*} \omega_{j,k} x_{k,i} \tag{2.92}$$

$$N_{m,i} = \frac{1}{1 + e^{n_{m,i}}} \tag{2.93}$$

$$\tilde{P}_{j,i} = \sum_{m=1}^{m^*} \gamma_{m,i} N_{j,i}, \quad \text{for } j = 1, 2, 3 \tag{2.94}$$

$$\sum_{m=1}^{m^*} \gamma_{m,i} = 1, \gamma_{m,i} \geq 0 \tag{2.95}$$

$$p_{j,i} = \frac{P_{j,i}}{\sum_{j=1}^{3} P_{j,i}} \tag{2.96}$$

The parameters of both the logistic and neural network models are estimated by maximizing a similar likelihood function:

$$\Lambda = \prod_{i=0}^{i=i^*} (p_{1,i})^{y_{1,i}} (p_{2,i})^{y_{2,i}} (p_{3,i})^{y_{3,i}} \tag{2.97}$$

The success of these alternative models is readily tabulated by the percentage of correct predictions for particular categories.

2.8   The Black Box Criticism and Data Mining

Like polynomial approximation, neural network estimation is often criticized as a *black box*. How do we justify the number of parameters, neurons, or hidden layers we use in a network? How does the design of the network relate to "priors" based on underlying economic or financial theory? Thomas Sargent (1997), quoting Lucas's advice to researchers, reminds us to beware of economists bearing "free parameters." By "free," we mean parameters that cannot be justified or restricted on theoretical grounds.

Clearly, models with a large number of parameters are more flexible than models with fewer parameters and can explain more variation in the data. But again, we should be wary. A criticism closely related to the black box issue is even more direct: a model that can explain everything, or nearly everything, in reality explains nothing. In short, models that are too good to be true usually are.

Of course, the same criticism can be made, *mutatis mutandis*, of linear models. All too often, the lag length of autoregressive models is adjusted to maximize the in-sample explanatory power or minimize the out-of-sample forecasting errors. It is often hard to relate the lag structure used in many linear empirical models to any theoretical priors based on the underlying optimizing behavior of economic agents.

Even more to the point, however, is the criticism of Wolkenhauer (2001): "formal models, if applicable to a larger class of processes are not specific (precise) enough for a particular problem, and if accurate for a particular problem they are usually not generally applicable" [Wolkenhauer (2001), p. xx].

The black box criticism comes from a desire to tie down empirical estimation with the underlying economic theory. Given the assumption that households, firms, and policy makers are rational, these agents or actors make decisions in the form of optimal feedback rules, derived from constrained dynamic optimization and/or strategic interaction with other players. The agents fully know their economic environment, and always act optimally or strategically in a fully rational manner.

The case for the use of neural networks comes from relaxing the assumption that agents fully know their environment. What if decision makers have to learn about their environment, about the nature of the shocks and underlying production, the policy objectives and feedback rules of the government, or the ways other players formulate their plans? It is not too hard to imagine that economic agents have to use approximations to capture and learn the way key variables interact in this type of environment.

From this perspective, the black box attack could be turned around. Should not fundamental theory take seriously the fact that economic decision makers are in the process of learning, of approximating their environment? Rather than being characterized as rational and all knowing, economic decision makers are boundedly rational and have to learn by working with several approximating models in volatile environments. This is what Granger and Jeon (2001) mean by "thick modeling."

Sargent (1999) himself has shown us how this can be done. In his book *The Conquest of American Inflation,* Sargent argues that inflation policy "emerges gradually from an adaptive process." He acknowledges that his "vindication" story "backs away slightly from rational expectations," in that policy makers used a 1960 Phillips curve model, but they "recurrently re-estimated a distributed lag Phillips curve and used it to reset a target inflation–unemployment rate pair" [Sargent (1999), pp. 4–5].

The point of Sargent's argument is that economists should model the actors or agents in their environments not as all-knowing rational angels who know the true model but rather in their own image and likeness, as econometricians who have to approximate, in a recursive or ongoing process, the complex interactions of variables affecting them. This book shows how one form of approximation of the complex interactions of variables affecting economic and financial decision makers takes place.

More broadly, however, there is the need to acknowledge model uncertainty in economic theory. As Hansen and Sargent (2000) point out, to say that a model is an approximation is to say that it approximates another model. Good theory need not work under the "communism of models," that the people being modeled "know the model" [Hansen and Sargent (2000), p. 1]. Instead, the agents must learn from a variety of models, even misspecified models.

Hansen and Sargent invoke the Ellsberg paradox to make this point. In this setup, originally put forward by Daniel Ellsberg (1961), there is a choice between two urns, one that contains 50 red balls and 50 black balls, and the second urn, in which the mix is unknown. The players can choose which urn to use and place bets on drawing red or black balls, with replacement. After a series of experiments, Ellsberg found that the first urn was more frequently chosen. He concluded that people behave in this way to avoid ambiguity or uncertainty. They prefer risk in which the probabilities are known to situations of uncertainty, when they are not.

However, Hansen and Sargent ask, when would we expect the second urn to be chosen? If the agents can learn from their experience over time, and readjust their erroneous prior subjective probabilities about the likelihood of drawing red or black from the second urn, there would be every reason to choose the second urn. Only if the subjective probabilities quickly converged to 50-50 would the players become indifferent. This simple example illustrates the need, as Hansen and Sargent contend, to model decision making in dynamic environments, with model approximation error and learning [Hansen and Sargent (2000), p. 6].

However, there is still the temptation to engage in data mining, to overfit a model by using increasingly complex approximation methods.

The discipline of Occam's razor still applies: simpler more transparent models should always be preferred over more complex less transparent approaches. In this research, we present simple neural network alternatives to the linear model and assess the performance of these alternatives by time-honored statistical criteria as well as the overall usefulness of these models for economic insight and decision making. In some cases, the simple linear model may be preferable to more complex alternatives; in others, neural network approaches or combinations of neural network and linear approaches clearly dominate. The point we wish to make in this research is that neural networks serve as a useful and readily available complement to linear methods for forecasting and empirical research relating to financial engineering.

2.9   Conclusion

This chapter has presented a variety of networks for forecasting, for dimensionality reduction, and for discrete choice or classification. All of these networks offer many options to the user, such as the selection of the number of hidden layers, the number of neurons or nodes in each hidden layer, and the choice of activation function with each neuron. While networks can easily get out of hand in terms of complexity, we show that the most useful network alternatives to the linear model, in terms of delivering improved performance, are the relatively simple networks, usually with only one hidden layer and at most two or three neurons in the hidden layer. The network alternatives never do worse, and sometimes do better, in the examples with artificial data (Chapter 5), with automobile production, corporate bond spreads, and inflation/deflation forecasting (Chapters 6 and 7).

Of course, for classification, the benchmark models are discriminant analysis, as well as nonlinear logit, probit, and Weibull methods. The neural network performs at least as well as or better than all of these more familiar methods for predicting default in credit cards and in banking-sector fragility (Chapter 8).

For dimensionality reduction, the race is between linear principal components and the neural net auto-associate mapping. We show, in the example with swap-option cap-floor volatility measures, that both methods are equally useful for in-sample power but that the network outperforms the linear methods for out-of-sample performance (Chapter 9).

The network architectures can mutate, of course. With a multilayer perceptron or feedforward network with several neurons in a hidden layer, it is always possible to specify alternative activation functions for the different neurons, with a logsigmoid function for one neuron, a tansig function for another, a cumulative Gaussian density for a third. But most researchers have found the "plain vanilla" multilayer perceptron network with logsigmoid activation functions fairly reliable and as accurate as more complex alternatives.

2.9.1   MATLAB Program Notes

The MATLAB program for estimating a multilayer perceptron or feedforward network on my webpage is the program *ffnet9.m* and uses the subfunction *ffnet9fun.m*. There are similar programs for recurrent Elman networks and jump connection networks: *ffnet9_elman.m*, *ffnet9fun_elman.m*, *ffnet9_jump.m*, and *ffnet9fun_jump.m*. The programs have instructions for the appropriate input arguments as well as descriptions of the outputs of the program.

For implementing a GARCH model, there is a program *mygarch.m*, which invokes functions supplied by the MATLAB Garch Toolbox.

For linear estimation, there is the *ols.m* program. This program has several subfunctions for diagnostics.

The classification models use the following programs: *classnet.m*, *classnetfun.m*, *logit.m*, *probit.m*, *gompit.m*.

For principal components, the programs to use are *nonlinpc.m* and *nonlinpcfun.m*. These functions in turn invoke the MATLAB program, *princomp.m*, which is part of the MATLAB Statistics Toolbox.

2.9.2   Suggested Exercises

For deriving the ridgelet network function, described in Section 2.4.4, you can use the MATLAB Symbolic Toolbox. It is easy to use and saves a lot of time and trouble. At the very least, in writing code, you can simply cut and paste the derivative formulae from this Toolbox to your own programs.

Simply type in the command *funtool.m*, and in the box beside "f=" type in the standard normal Gaussian formula, "inv(2 *pi) * exp($-$x ^2)" (no need for parentheses). Then click on the derivative button, "df/dx," five times until you arrive at the formula given for the ridgelet network.

Repeat the above exercise for the logsigmoid function, setting in the formula next of "f=" "inv(1+exp($-$x))". After taking the derivatives a number of times, compare the graph of the function, in the interval $[-2\ pi, 2\ pi]$ with that of the corresponding $(n-1)$ derivative of the Gaussian function. Why do they start to look alike?

3

Estimation of a Network with Evolutionary Computation

If the specification of the neural network for approximation appears to be inspired by biology, the reader will no doubt suspect that the best way to estimate or train a network is inspired by genetics and evolution. Estimating a nonlinear model is always tricky business. The programs may fail to converge, or they may converge to locally, rather than globally, optimal estimates. We show that the best way to estimate a network, to implement the network, is to harness the power of evolutionary genetic search algorithms.

3.1 Data Preprocessing

Before moving to the actual estimation, however, the first order of business is to adjust or scale the data and to remove nonstationarity. In other words, the first task is data preprocessing. While linear models also require that data be stationary and seasonally adjusted, scaling is critically important for nonlinear estimation, since such scaling reduces the search space for finding the optimal coefficient estimates.

3.1.1 Stationarity: Dickey-Fuller Test

Before starting work with any time series as a dependent variable, we must ensure that the data represent *covariance stationary* time series.

This means that the first and second moments — means, variances, and covariances — are constant through time. Since statistical inference is based on the assumption of fixed means, variances, and covariances, it is essential to ensure that the variables in question are indeed stationary.

The most commonly used test is the one proposed by Dickey and Fuller (1979), for a given series $\{y_t\}$:

$$\Delta y_t = \rho y_{t-1} + \alpha_1 \Delta y_{t-1} + \alpha_2 \Delta y_{t-2} + \cdots + \alpha_k \Delta y_{t-k} + \varepsilon_t \tag{3.1}$$

where $\Delta y_t = y_t - y_{t-1}$, $\rho, \alpha_1, \ldots, \alpha_k$ are coefficients to be estimated, and ε_t is a random disturbance term with mean zero and constant variance. Thus, $\mathbf{E}(\varepsilon_t) = 0$, and $\mathbf{E}(\varepsilon_t^2) = \sigma^2$.

The null hypothesis under this test is $\rho = 0$. In this case, the regression model reduces to the following expression:

$$y_t = y_{t-1} + \alpha_1 \Delta y_{t-1} + \alpha_2 \Delta y_{t-2} + \cdots + \alpha_k \Delta y_{t-k} + \varepsilon_t \tag{3.2}$$

Under this null hypothesis, y_t at any moment will be equal to y_{t-1} plus or minus the effect of the terms given by the sum of $\alpha_i \Delta y_{t-i}$, $i = 1, \ldots, k$. In this case, the long-run expected value of the series, when $y_t = y_{t-1}$, becomes indeterminate. Or perhaps more succinctly, the mean at any given time is conditional on past values of y_t. With $\rho = 0$, the series is called *nonstationary*, or a *unit root process*.

The relevant alternative hypothesis is $\rho < 0$. With $\rho = -1$, the model reduces to the following expression:

$$y_t = \alpha_1 \Delta y_{t-1} + \alpha_2 \Delta y_{t-2} + \cdots + \alpha_k \Delta y_{t-k} + \varepsilon_t \tag{3.3}$$

In the long run, with $y_t = y_{t-1}$, by definition, $\Delta y_{t-i} = 0$, for $i = i, \ldots, k$, so that the expected value of y_t, $\mathbf{E}y_t = \mathbf{E}(\varepsilon_t) = 0$.

If there is some persistence in the model, with ρ falling in the interval between $[-1, 0]$, the relevant regression becomes:

$$y_t = (1 + \rho)y_{t-1} + \alpha_1 \Delta y_{t-1} + \alpha_2 \Delta y_{t-2} + \cdots + \alpha_k \Delta y_{t-k} + \varepsilon_t \tag{3.4}$$

In this case, in the long run, with $y_t = y_{t-1}$, it is still true that $\Delta y_{t-i} = 0$, for $i = i, \ldots, k$. The only difference is that the expression for the long-run mean reduces to the following expression, with $\rho^* = (1 + \rho)$:

$$y_t(1 - \rho^*) = \varepsilon_t \tag{3.5}$$

In this case, the expected value of y_t, $\mathbf{E}y_t$, is equal to $\mathbf{E}(\varepsilon_t)/(1 - \rho^*)$.

It is thus crucial to ensure that the coefficient ρ is significantly less than zero for stationarity. The tests of Dickey and Fuller are essentially modified, one-sided t-tests of the hypothesis of $\rho < 0$ in a linear regression.

Augmented Dickey-Fuller tests allow the presence of constant and trend terms in the preceding regressions.

The stationarity tests of Dickey and Fuller led to the development of the Phillips and Perron (1988) test. This test goes beyond Dickey and Fuller in that it permits a joint test of significance of the coefficients of the autoregressive term as well as the trend and constant terms.[1] Further work on stationarity has involved tests for structural breaks in univariate nonstationarity time series [see, for example, Benerjee, Lumsdaine, and Stock (1992); Lumsdaine and Papell (1997); Perron (1989); and Zivot and Andrews (1992)].

Fortunately, for most financial time-series data such as share prices, nominal money supply, and gross domestic product, logarithmic first differencing usually transforms these nonstationarity time series into stationarity series. Logarithmic first differencing simply involves taking the logarithmic value of a series Z, and then taking its first difference.

$$\Delta z_t = \ln(Z_t) - \ln(Z_{t-1}) \tag{3.6}$$

$$z_t \equiv \ln(Z_t) \tag{3.7}$$

3.1.2   Seasonal Adjustment: Correction for Calendar Effects

A further problem with time-series data arises from seasonal or calendar effects. With quarterly or monthly data, there are obvious end-of-year December spikes in consumer spending. With daily data, there are effects associated with particular months, days of the week, and holidays. The danger of not adjusting the data for these seasonal factors in nonlinear neural network estimation is overfitting the data. The nonlinear estimation process will continue to fine tune the fitting of the model or look for needlessly complex representations to account for purely seasonal factors.

Of course, the danger of any form of seasonal adjustment is that one may extract useful information from the data. It is thus advisable to work with the raw, seasonally unadjusted series as a benchmark.

Fortunately, for quarterly or monthly data, one may use a simple dummy variable regression method. For quarterly data, for example, one estimates the following regression:

$$\Delta z = Q'\beta + u \tag{3.8}$$

---

[1]See Hamilton (1994), Chapter 17, for a detailed discussion of unit roots and tests for stationarity in time series.

where Δz_t is the stationarity raw series, the matrix $Q = [Q_2, Q_3, Q_4]$ represents dummy variables for the second, third, and fourth quarters of the year, and u is the residual, or everything in the raw series that cannot be explained by the quarterly dummy variables. These dummy variables take on values of 1 when the observation falls in the respective quarter, and zero otherwise.

A similar procedure is performed for monthly data, with eleven monthly dummy variables.[2]

For daily data, the seasonal filtering regression is more complicated. Gallant, Rossi, and Tauchen (1992) propose the following sets of regressors:

1. Day-of-week dummies for Tuesday through Friday

2. Dummy variables for each of the number of nontrading days preceding the current trading day[3]

3. Dummy variables for the months of March, April, May, June, July, August, September, October, and November

4. Dummy variables for each week of December and January

In the Gallant-Rossi-Tauchen procedure, one first regresses the stationarity variable Δz_t on the set of adjustment variables A_t, where A is the matrix of dummy variables, for days of the week, months, weeks in December and January, and the number of nontrading days preceding the current trading day:

$$\Delta z = A'\beta + u \tag{3.9}$$

Gallant, Rossi, and Tauchen also allow the variance, as well as the mean, of the data, to be adjusted for the calendar effects. One simply does a regression of the logarithm of $\widehat{u}^2$ on the set of dummy calendar variables, A, and the trend terms $[t\ t^2]$, where $t = 1, 2, \ldots, T$, with T representing the number of observations. The regression equation becomes:

$$\ln(\widehat{u}^2) = \underline{A}'\gamma + \epsilon \tag{3.10}$$

$$\underline{A} = [A\ t\ t^2] \tag{3.11}$$

---

[2]In both cases, omit one dummy variable to avoid collinearity with the constant term in the regressions.

[3]Fortunately, most financial websites have information on holidays in most countries, so that one may obtain the relevant data for the number of nontrading days preceding each date.

TABLE 3.1. Gallant-Rossi-Tauchen Procedure for Calendar Adjustment

Step	Operation
Define and quantify calendar dummy matrix, A	$[A]$
Regress dependent variables on dummy matrix	$\Delta z = A'\beta + u$
Form expanded dummy matrix	$\underline{A} = [A \ t \ t^2]$
Regress squared residuals on expanded matrix	$\ln(\widehat{u}^2) = \underline{A}'\gamma + \epsilon$
Transform residuals u to Δz^*	$\Delta z^* = a + b\left[\dfrac{\widehat{u}}{\exp\left(\frac{\underline{A}'\gamma}{2}\right)}\right]$

They also propose a final linear transformation so that the adjusted series Δz^* has the same sample mean and variances as the original raw series:

$$\Delta z^* = a + b\left[\frac{\widehat{u}}{\exp\left(\frac{\underline{A}'\gamma}{2}\right)}\right] \tag{3.12}$$

with the variables a and b chosen to ensure that the sample means and variances of the two series are identical.

Table 3.1 summarizes the Gallant-Rossi-Tauchen procedure for calendar adjustment.

Of course, seasonal adjustment is also done through smoothing of the original data series, usually through moving average filters. Many series available in national income accounts in fact are already seasonally adjusted by such smoothing methods.

The advantages of different seasonal adjustment procedures depends on the goal of the research. If the focus is on reliable parameter estimates of an econometric model, the dummy variable approach is superior.[4] In all of this calendar adjustment, we are replacing the original series with artificially adjusted data. There may be resistance by decision makers to this approach, for example, in options pricing, if the underlying adjusted return series does not match closely the actual observed return series. For this reason, it is a good strategy to examine the results of the models with the actual and the calendar-adjusted series. We would expect greater precision with the adjusted series, and quicker convergence, but the overall results should not be drastically different.

---

[4]See Beck (1981) for a discussion of different types of seasonal adjustment for econometric model estimation.

3.1.3 Data Scaling

When input variables $\{x_t\}$ and stationary output variables $\{y_t\}$ are used in a neural network, preprocessing or scaling facilitates the nonlinear estimation process. The reason scaling is helpful, even crucial, is that the use of very high or low numbers, or series with a few very high or very low outliers, can cause underflow or overflow problems, with the computer stopping, or as Judd [(1998), p. 99] points out, the computer continuing by assigning a value of zero to the values being minimized.

When logsigmoid or tansigmoid neurons are used, to be sure, scaling is a necessary step. If the data are not scaled to a reasonable interval, such as $[0, 1]$ or $[-1, 1]$, then the neurons will set reasonably large values simply at 1, and reasonably low values at 0 (for logsigmoid neurons) or -1 (for tansig neurons). Without scaling, a great deal of information from the data is likely to be lost, since the neurons will simply transmit values of minus one, zero, or plus one for many values of the input data.

There are two main numeric ranges the network specialists use in *linear scaling functions*: zero to one, denoted $[0, 1]$, and minus one to plus one denoted by $[-1, 1]$.

Linear scaling functions make use of the maximum and minimum values of the series $[y\ x]$. The linear scaling function for zero to one transforms a variable x_k into x_k^* in the following way:

$$x_{k,t}^* = \frac{x_{k,t} - \min(x_k)}{\max(x_k) - \min(x_k)} \tag{3.13}$$

The linear scaling function for $[-1, 1]$, transforming a variable x_k into x_k^{**}, has the following form:

$$x_{k,t}^{**} = 2 \cdot \frac{x_{k,t} - \min(x_k)}{\max(x_k) - \min(x_k)} - 1 \tag{3.14}$$

A nonlinear scaling method proposed by Dr. Helge Petersohn of the University of Leipzig, transforming a variable x_k to z_k, allows one to specify the range $0 < z_{k,t} < 1$, or $\langle 0, 1 \rangle$. The Petersohn scaling function works in the following way:

$$z_{k,t} = \frac{1}{1 + \exp\left[\left(\dfrac{\ln[\bar{z}_k^{-1} - 1] - \ln[\underline{z}_k^{-1} - 1]}{\max(x_k) - \min(x_k)}\right)[x_{k,t} - \min(x_k)] + \ln[\underline{z}_k^{-1} - 1]^{-1}\right]} \tag{3.15}$$

Finally, James DeLeo of the National Institutes of Health suggests scaling the data in a two-step procedure: first, standardizing a series x, to obtain z, and then taking the logsigmoid transformation of the standardized series z:

$$x^* = \frac{1}{1 + \exp(-z)} \tag{3.16}$$

$$z = \frac{x - \overline{x}}{\sigma_x} \tag{3.17}$$

Which type of scaling function works best depends on the quality of the results. There is no way to decide which scaling function works best, on *a priori* grounds, given the features of the data. The best strategy is to estimate the model with different types of scaling functions to find out which one gives the best performance, based on in-sample criteria discussed in the following section.

3.2 The Nonlinear Estimation Problem

Finding the coefficient values for a neural network, or any nonlinear model, is not an easy job — certainly not as easy as parameter estimation with a linear approximation. A neural network is a highly complex nonlinear system. There may be a multiplicity of locally optimal solutions, none of which deliver the best solution in terms of minimizing the differences between the model predictions $\widehat{y}$ and the actual values of y. Thus, neural network estimation takes time and involves the use of alternative methods.

Briefly, in any nonlinear system, we need to start the estimation process with initial conditions, or guesses of the parameter values we wish to estimate. Unfortunately, some guesses may be better than others for moving the estimation process to the best coefficients for the optimal forecast. Some guesses may lead us to a local optimum, that is, the best forecast in the neighborhood of the initial guess, but not the coefficients for giving the best forecast if we look a bit further afield from the initial guesses for the coefficients.

Figure 3.1 illustrates the problem of finding globally optimal or globally minimal points on a highly nonlinear surface.

As Figure 3.1 shows, an initial set of weight values anywhere on the x axis may lie near to a local or global maximum rather than a minimum, or near to a saddle point. A minimum or maximum point has a slope or derivative equal to zero. At a maximum point, the second derivative, or change in the slope, is negative, while at a minimum point, the change in the slope is positive. At a saddle point, both the slope and the change in the slope are zero.

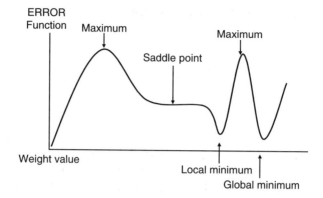

FIGURE 3.1. Weight values and error function

   As the weights are adjusted, one can get stuck at any of the many positions where the derivative is zero, or the curve has a flat slope. Too large an
adjustment in the learning parameter may bring one's weight values from
a near-global minimum point to a maximum or to a saddle point. However,
too small an adjustment may keep one trapped near a saddle point for
quite some time during the training period.
   Unfortunately, there is no silver bullet for avoiding the problems of local
minima in nonlinear estimation. There are only strategies involving re-
estimation or stochastic evolutionary search.
   For finding the set of coefficients or weights $\Omega = \{\omega_{k,i}, \gamma_k\}$ in a network
with a single hidden layer, or $\Omega = \{\omega_{k,i}, \rho_{l,k}, \gamma_l\}$ in a network with two
hidden layers, we minimize the loss function Ψ, defined again as the sum of
squared differences between the actual observed output y and $\widehat{y}$, the output
predicted by the network:

$$\min_{\Omega} \Psi(\Omega) = \sum_{t=1}^{T} (y_t - \widehat{y}_t)^2 \tag{3.18}$$

$$\widehat{y}_t = f(x_t; \Omega) \tag{3.19}$$

where T is the number of observations of the output vector y, and $f(x_t; \Omega)$
is a representation of the neural network.
   Clearly, $\Psi(\Omega)$ is a nonlinear function of Ω. All nonlinear optimization
starts with an initial guess of the solution, Ω_0, and searches for better solutions, until finding the best possible solution within a reasonable amount
of searching.

We discuss three ways to minimize the function $\Psi(\Omega)$:

1. A local gradient-based search, in which we compute first- and second-order derivatives of Ψ with respect to elements of the parameter vector Ω, and continue with updating of the initial guess of Ω, by derivatives, until stopping criteria are reached

2. A stochastic search, called simulated annealing, which does not rely on the use of first- and second-order derivatives, but starts with an initial guess Ω_0, and proceeds with random updating of the initial coefficients until a "cooling temperature" or stopping criterion is reached

3. An evolutionary stochastic search, called the genetic algorithm, which starts with a population of p initial guesses, $[\Omega_{01}, \Omega_{02} \ldots \Omega_{0p}]$, and updates the population of guesses by genetic selection, breeding, and mutation, for many generations, until the best coefficient vector is found among the last-generation population

All of this discussion is rather straightforward for students of computer science or engineering. Those not interested in the precise details of nonlinear optimization may skip the next three subsections without fear of losing their way in succeeding sections.

3.2.1 Local Gradient-Based Search: The Quasi-Newton Method and Backpropagation

To minimize any nonlinear function, we usually begin by initializing the parameter vector Ω at any initial value, Ω_0, perhaps at randomly chosen values. We then iterate on the coefficient set Ω until Ψ is minimized, by making use of first- and second-order derivatives of the error metric Ψ with respect to the parameters. This type of search, called a gradient-based search, is for the optimum in the neighborhood of the initial parameter vector, Ω_0. For this reason, this type of search is a local search.

The usual way to do this iteration is through the quasi-Newton algorithm. Starting with the initial set of the sum of squared errors, $\Psi(\Omega_0)$, based on the initial coefficient vector Ω_0, a second-order Taylor expansion is used to find $\Psi(\Omega_1)$:

$$\Psi(\Omega_1) = \Psi(\Omega_0) + \nabla_0(\Omega_1 - \Omega_0) + .5(\Omega_1 - \Omega_0)' H_0(\Omega_1 - \Omega_0) \qquad (3.20)$$

where ∇_0 is the gradient of the error function with respect to the parameter set Ω_0 and H_0 is the Hessian of the error function.

Letting $\Omega_0 = [\Omega_{0,1}, \ldots, \Omega_{0,k}]$, be the initial set of k parameters used in the network, the gradient vector ∇_0 is defined as follows:

$$\nabla_0 = \begin{pmatrix} \frac{\Psi(\Omega_{0,1}+h_1,\ldots,\Omega_{0,k})-\Psi(\Omega_{0,1},\ldots,\Omega_{0,k})}{h_1} \\ \frac{\Psi(\Omega_{0,1},\ldots,\Omega_{0,i}+h_i,\ldots,\Omega_{0,k})-\Psi(\Omega_{0,1},\ldots,\Omega_{0,k})}{h_i} \\ . \\ . \\ \frac{\Psi(\Omega_{0,1},\ldots,\Omega_{0,i},\ldots,\Omega_{0,k}+h_k)-\Psi(\Omega_{0,1},\ldots,\Omega_{0,k})}{h_k} \end{pmatrix} \tag{3.21}$$

The denominator h_i is usually set at $\max(\epsilon, \epsilon\Omega_{0,i})$, with $\epsilon = 10^{-6}$.

The Hessian H_0 is the matrix of second-order partial derivatives of Ψ with respect to the elements of Ω_0, and is computed in a similar manner as the Jacobian or gradient vector. The cross-partials or off-diagonal elements of the matrix H_0 are given by the formula:

$$\frac{\partial^2 \Psi}{\partial \Omega_{0,i} \partial \Omega_{0,j}} = \frac{1}{h_j h_i}$$

$$\times \begin{bmatrix} \{\Psi(\Omega_{0,1},\ldots,\Omega_{0,i}+h_i,\Omega_{0,j}+h_j,\ldots,\Omega_{0,k})-\Psi(\Omega_{0,1},\ldots,\Omega_{0,i},\ldots,\Omega_{0,j}+h_j,\ldots,\Omega_{0,k})\} \\ -\{\Psi(\Omega_{0,1},\ldots,\Omega_{0,i}+h_i,\Omega_{0,j},\ldots,\Omega_{0,k})-\Psi(\Omega_{0,1},\ldots,\Omega_{0,k})\} \end{bmatrix}$$

$$\tag{3.22}$$

while the direct second-order partials or diagonal elements are given by:

$$\frac{\partial^2 \Psi}{\partial \Omega_{0,i}^2} = \frac{1}{h_i^2} \left(\begin{array}{c} \Psi(\Omega_{0,1},\ldots,\Omega_{0,i}+h_i,\ldots,\Omega_{0,k}) - 2\Psi(\Omega_{0,1},\ldots,\Omega_{0,k}) \\ +\Psi(\Omega_{0,1},\ldots,\Omega_{0,i}-h_i,\ldots,\Omega_{0,k}) \end{array} \right)$$

$$\tag{3.23}$$

To find the direction of a change of the parameter set from iteration 0 to iteration 1, one simply minimizes the error function $\Psi(\Omega_1)$ with respect to $(\Omega_1 - \Omega_0)$. The following formula gives the evolution of the parameter set Ω from the initial specification at iteration 0 to its value at iteration 1.

$$(\Omega_1 - \Omega_0) = -H_0^{-1}\nabla_0 \tag{3.24}$$

The algorithm continues in this way, from iteration 1 to 2, 2 to 3, $n-1$ to n, until the error function is minimized. One can set a tolerance criterion, stopping when there are no further changes in the error function below a given tolerance value. Alternatively, one may simply stop when a specified maximum number of iterations is reached.

The major problem with this method, as in any nonlinear optimization method, is that one may find local rather than global solutions, or a saddle-point solution for the vector Ω^*, which minimizes the error function.

Where the algorithm ends in the optimization process crucially depends on the choice of the initial parameter vector Ω_0. The most commonly used approach is to start with one random vector, iterate until convergence is achieved, and begin again with another random parameter vector, iterate until converge, and compare the final results with the initial iteration. Another strategy is to repeat this minimization many times until it reaches a potential global minimum value over the set of minimum values.

Another problem is that as iterations progress, the Hessian matrix H at iteration n^* may also become nonsingular, so that it is impossible to obtain $H_{n^*}^{-1}$ at iteration n^*. Commonly used numerical optimization methods approximate the Hessian matrix at various iteration periods.

The BFGS (Boyden-Fletcher-Goldfarb-Shanno) algorithm approximates H_n^{-1} at step n on the basis of the size of the change in the gradient ∇_n-∇_{n-1} relative to the change in the parameters $\Omega_n - \Omega_{n-1}$. Other algorithms available are the Davidon-Fletcher-Powell (D-F-P) and Berndt, Hall, Hall, and Hausman (BHHH). [See Hamilton (1994), p. 139.]

All of these approximation methods frequently blow up when there are large numbers of parameters or if the functional form of the neural network is sufficiently complex. Paul John Werbos (1994) first developed the backpropagation method in the 1970s as an alternative for estimating neural network coefficients under gradient-search. Backpropagation is a very manageable way to estimate a network without having to iterate and invert the Hessian matrices under the BFGS, DFP, and BHHH routines. It remains the most widely used method for estimating neural networks. In this method, the inverse Hessian matrix, $-H_0^{-1}$, is replaced by an identity matrix, with its dimension equal to the number of coefficients, k, multiplied by a learning parameter, ρ:

$$(\Omega_1 - \Omega_0) = -H_0^{-1}\nabla_0 \tag{3.25}$$

$$= -\rho \cdot \nabla_0 \tag{3.26}$$

Usually, the learning parameter ρ is specified at the start of the estimation, usually at small values, in the interval $[.05, .5]$, to avoid oscillations. The learning parameters can be endogenous, taking on different values as the estimation process appears to converge, when the gradients become smaller. Extensions of the backpropagation method allow different learning rates for different parameters. However, efficient as backpropagation may be, it still suffers from the trap of local rather than global minima, or saddle point convergence. Moreover, while low values of the learning parameters avoid oscillations, they may needlessly prolong the convergence process.

One solution for speeding up the process of backpropagation toward convergence is to add a momentum term to the above process, after a period of n training periods:

$$(\Omega_n - \Omega_{n-1}) = -\rho \cdot \nabla_{n-1} + \mu(\Omega_{n-1} - \Omega_{n-2}) \qquad (3.27)$$

The effect of adding the moment effect, with μ usually set to .9, is to enable the adjustment of the coefficients to roll or move more quickly over a plateau in the "error surface" [Essenreiter (1996)].

3.2.2 Stochastic Search: Simulated Annealing

In neural network estimation, where there are a relatively large number of parameters, Newton-based algorithms are less likely to be useful. It is difficult to invert the Hessian matrices in this case. Similarly, the initial parameter vector may not be in the neighborhood of the best solution, so a local search may not be very efficient.

An alternative search method for optimization is simulated annealing. It does not require taking first- or second-order derivatives. Rather, it is a stochastic search method. Originally due to Metropolis et al. (1953), later developed by Kirkpatrick, Gelatt, and Vecchi (1983), it originates from the theory of statistical mechanics. According to Sundermann (1996), this method is based on the analogy between the annealing of solids and solving optimization.

The simulated annealing process is described in Table 3.2. The basic message of this approach is well summarized by Haykin (1994): "when optimizing a very large and complex system (i.e. a system with many degrees of freedom), instead of always going downhill, try to go downhill most of the time" [Haykin (1994), p. 315].

As Table 3.2 shows, we again start with a candidate solution vector, Ω_0, and the associated error criterion, Ψ_0. A shock to the solution vector is then randomly generated, Ω_1, and we calculate the associated error metric, Ψ_1. We always accept the new solution vector if the error metric decreases. However, since the initial guess Ω_0 may not be very good, there is a small chance that the new vector, even if it does not reduce the error metric, may be moving in the right direction to a more global solution. So with a probability $P(j)$, conditioned by the Metropolis ratio $M(j)$, the new vector may be accepted, even though the error metric actually increases. The rationale for accepting a new vector Ω_i even if the error Ψ_i is greater than Ψ_{i-1}, is to avoid the pitfall of being trapped in a local minimum point. This allows us to search over a wider set of possibilities.

As Robinson (1995) points out, simulated annealing consists of running the accept/reject algorithm between the temperature extremes. Many changes are proposed, starting at the high temperatures, which explore the parameter space. With gradually decreasing temperature, however, the

TABLE 3.2. Simulated Annealing for Local Optimization

Definition	Operation
Specify temperature and cooling schedule parameter $\overline{T}$	$T(j) = \dfrac{\overline{T}}{1 + \ln(j)}$
Start random process at j = 0, continue till j = (1,2,...,$\overline{T}$)	
Initialize solution vector and error metric	Ω_0, Ψ_0
Randomly perturbate solution vector, obtain error metric	$\widehat{\Omega}_j, \widehat{\Psi}_j$
Generate P(j) from uniform distribution	$0 \leq P(j) \leq 1$
Compute metropolis ratio M(j)	$M(j) = \exp\left[\dfrac{-\left(\widehat{\Psi}_j - \Psi_{j-1}\right)}{T(j)} \right]$
Accept new vector $\Omega_j = \widehat{\Omega}_j$ unconditionally	$\Omega_j = \widehat{\Omega}_j \Leftrightarrow \left(\widehat{\Psi}_j - \Psi_{j-1}\right) < 0$
Accept new vector $\Omega_j = \widehat{\Omega}_j$ conditionally	$P(j) \leq M(j)$
Continue process till j = $\overline{T}$	

algorithm becomes "greedy." As the temperature $T(j)$ cools, changes are more and more likely to be accepted only if the error metric decreases.

To be sure, simulated annealing is not strictly a global search. Rather it is a random search for helping to escape a likely local minimum and move to a better minimum point. So it is best used after we have converged to a given point, to see if there are better minimum points in the neighborhood of the initial minimum.

As we see in Table 3.2, the current state of the system, or coefficient vector $\widehat{\Omega}_j$, depends only on the previous state $\widehat{\Omega}_{j-1}$, and a transition probability $P(j - 1)$ and is thus independent of all previous outcomes. We say that such a system has the Markov chain property. As Haykin (1994) notes, an important property of this system is asymptotic convergence, for which Geman and Geman (1984) gave us a mathematical proof. Their theorem, summarized from Haykin (1994, p. 317), states the following:

Theorem 1 *If the temperature* T(k) *employed in executing the k-th step satisfies the bound* T(k) $\geq \overline{T}$/ log(1+k) *for every k, where* $\overline{T}$ *is a sufficiently large constant independent of* k, *then with probability 1 the system will converge to the minimum configuration.*

A similar theorem has been derived by Aarts and Korst (1989). Unfortunately, the annealing schedule given in the preceding theorem would be extremely slow — much too slow for practical use. When we resort to finite-time approximation of the asymptotic convergence properties, we are no longer guaranteed that we will find the global optimum with probability one.

For implementing the algorithm in finite-time approximation, we have to decide on the key parameters in the annealing schedule. Van Laarhoven and Aarts (1988) have developed more detailed annealing schedules than the one presented in Table 3.2. Kirkpatrick, Gelatt, and Vecchi (1983) offered suggestions for the starting temperature $\overline{T}$ (it should be high enough to ensure that all proposed transitions are accepted by algorithm), a linear alternative for the temperature decrement function, with $T(k) = \alpha T(k-1), .8 \leq \alpha \leq .99$, as well as a stopping rule (the system is "frozen" if the desired number of acceptances is not achieved at three successive temperatures). Adaptive simulated annealing is a further development which has proven to be faster and has become more widely used [Ingber (1989)].

3.2.3   Evolutionary Stochastic Search: The Genetic Algorithm

Both the Newton-based optimization (including backpropagation) and simulated annealing (SA) start with one random initialization vector Ω_0. It should be clear that the usefulness of both of these approaches to optimization crucially depends on how good this initial parameter guess really is. The genetic algorithm or GA helps us come up with a better guess for using either of these search processes.

The GA reduces the likelihood of landing in a local minimum. We no longer have to approximate the Hessians. Like simulated annealing, it is a statistical search process, but it goes beyond SA, since it is an *evolutionary search process*.

The GA proceeds in the following steps.

Population Creation

This method starts not with one random coefficient vector Ω, but with a population N^* (an even number) of random vectors. Letting p be the size of each column vector, representing the total number of coefficients to be estimated in the neural network, we create a population N^* of p by 1 random vector.

$$\begin{pmatrix} \Omega_1 \\ \Omega_2 \\ \Omega_3 \\ \cdot \\ \cdot \\ \Omega_p \end{pmatrix}_1 \begin{pmatrix} \Omega_1 \\ \Omega_2 \\ \Omega_3 \\ \cdot \\ \cdot \\ \Omega_p \end{pmatrix}_2 \begin{pmatrix} \Omega_1 \\ \Omega_2 \\ \Omega_3 \\ \cdot \\ \cdot \\ \Omega_p \end{pmatrix}_i \cdots \begin{pmatrix} \Omega_1 \\ \Omega_2 \\ \Omega_3 \\ \cdot \\ \cdot \\ \Omega_p \end{pmatrix}_{N*} \tag{3.28}$$

Selection

The next step is to select two pairs of coefficients from the population at random, with replacement. Evaluate the fitness of these four coefficient vectors, in two pair-wise combinations, according to the sum of squared error function. Coefficient vectors that come closer to minimizing the sum of squared errors receive better fitness values.

This is a simple fitness tournament between the two pairs of vectors: the winner of each tournament is the vector with the best fitness. These two winning vectors (i, j) are retained for "breeding" purposes. While not always used, it has proven to be extremely useful for speeding up the convergence of the genetic search process.

$$\begin{pmatrix} \Omega_1 \\ \Omega_2 \\ \Omega_3 \\ \cdot \\ \cdot \\ \Omega_p \end{pmatrix}_i \begin{pmatrix} \Omega_1 \\ \Omega_2 \\ \Omega_3 \\ \cdot \\ \cdot \\ \Omega_p \end{pmatrix}_j$$

Crossover

The next step is crossover, in which the two parents "breed" two children. The algorithm allows crossover to be performed on each pair of coefficient vectors i and j, with a fixed probability $p > 0$. If crossover is to be performed, the algorithm uses one of three difference crossover operations, with each method having an equal $(1/3)$ probability of being chosen:

1. *Shuffle crossover.* For each pair of vectors, k random draws are made from a binomial distribution. If the kth draw is equal to 1, the coefficients $\Omega_{i,p}$ and $\Omega_{j,p}$ are swapped; otherwise, no change is made.

2. *Arithmetic crossover.* For each pair of vectors, a random number is chosen, $\omega \in (0, 1)$. This number is used to create two new parameter vectors that are linear combinations of the two parent factors, $\omega \Omega_{i,p} + (1 - \omega)\Omega_{j,p}, (1 - \omega \Omega_{i,p} + \omega)\Omega_{j,p}$.

3. *Single-point crossover.* For each pair of vectors, an integer I is randomly chosen from the set $[1, k - 1]$. The two vectors are then cut at integer I and the coefficients to the right of this cut point, $\Omega_{i,I+1}, \Omega_{j,I+1}$ are swapped.

In binary-encoded genetic algorithms, single-point crossover is the standard method. There is no consensus in the genetic algorithm literature on which method is best for real-valued encoding.

Following the crossover operation, each pair of parent vectors is associated with two children coefficient vectors, which are denoted $C1(i)$ and $C2(j)$. If crossover has been applied to the pair of parents, the children vectors will generally differ from the parent vectors.

Mutation

The fifth step is mutation of the children. With some small probability $\widetilde{pr}$, which decreases over time, each element or coefficient of the two children's vectors is subjected to a mutation. The probability of each element is subject to mutation in generation $G = 1, 2, \ldots, G^*$, given by the probability $\widetilde{pr} = .15 + .33/G$.

If mutation is to be performed on a vector element, we use the following nonuniform mutation operation, due to Michalewicz (1996).

Begin by randomly drawing two real numbers r_1 and r_2 from the $[0, 1]$ interval and one random number s from a standard normal distribution. The mutated coefficient $\widetilde{\Omega}_{i,p}$ is given by the following formula:

$$\widetilde{\Omega}_{i,p} = \left\{ \begin{array}{l} \Omega_{i,p} + s[1 - r_2^{(1-G/G^*)^b}] \text{ if } r_1 > .5 \\ \Omega_{i,p} - s[1 - r_2^{(1-G/G^*)^b}] \text{ if } r_1 \leq .5 \end{array} \right\} \tag{3.29}$$

where G is the generation number, G^* is the maximum number of generations, and b is a parameter that governs the degree to which the mutation operation is nonuniform. Usually we set $b = 2$. Note that the probability of creating via mutation a new coefficient that is far from the current coefficient value diminishes as $G \rightarrow G^*$, where G^* is the number of generations. Thus, the mutation probability itself evolves through time.

The mutation operation is nonuniform since, over time, the algorithm is sampling increasingly more intensively in a neighborhood of the existing coefficient values. This more localized search allows for some fine tuning of the coefficient vector in the later stages of the search, when the vectors should be approaching close to a global optimum.

Election Tournament

The last step is the election tournament. Following the mutation operation, the four members of the "family" $(P1, P2, C1, C2)$ engage in a fitness tournament. The children are evaluated by the same fitness criterion used to evaluate the parents. The two vectors with the best fitness, whether parents or children, survive and pass to the next generation, while the two with the worst fitness value are extinguished. This election operator is due to Arifovic (1996). She notes that this election operator "endogenously controls the realized rate of mutation" in the genetic search process [Arifovic (1996), p. 525].

We repeat the above process, with parents i and j returning to the population pool for possible selection again, until the next generation is populated by N^* vectors.

Elitism

Once the next generation is populated, we can introduce elitism (or not). Evaluate all the members of the new generation and the past generation according to the fitness criterion. If the best member of the older generation dominated the best member of the new generation, then this member displaces the worst member of the new generation and is thus eligible for selection in the coming generation.

Convergence

One continues this process for G^* generations. Unfortunately, the literature gives us little guidance about selecting a value for G^*. Since we evaluate convergence by the fitness value of the best member of each generation, G^* should be large enough so that we see no changes in the fitness values of the best for several generations.

3.2.4 Evolutionary Genetic Algorithms

Just as the genetic algorithm is an evolutionary search process for finding the best coefficient set Ω of p elements, the parameters of the genetic algorithm, such as population size, probability of crossover, initial mutation probability, use of elitism or not, can evolve themselves. As Michalewicz and Fogel (2002) observe, "let's admit that finding good parameter values for an evolutionary algorithm is a poorly structured, ill-defined, complex problem. But these are the kinds of problems for which evolutionary algorithms are themselves quite adept" [Michalewicz and Fogel (2002), p. 281]. They suggest two ways to make a genetic algorithm evolutionary. One, as we suggested with the mutation probability, is to use a feedback rule from the state of the system which modifies a parameter during the search process. Alternatively, we can incorporate the training parameters into the solution by modifying Ω to include additional elements such as population size, use of elitism, or crossover probability. These parameters thus become subject to evolutionary search along with the solution set Ω itself.

3.2.5 Hybridization: Coupling Gradient-Descent, Stochastic, and Genetic Search Methods

The gradient-descent methods are the most commonly used optimization methods in nonlinear estimation. However, as previously noted, there is a strong danger of getting stuck in a local rather than a global minimum for a vector w, or in a saddlepoint. Furthermore, if using a Newton algorithm, the Hessian matrix may fail to invert, or become "near-singular," leading to imprecise or even absurd results for the coefficient vector of the neural network. When there are a large number of parameters, the statistically based simulated annealing search is a good alternative.

The genetic algorithm does not involve taking gradients or second derivatives and is a global and evolutionary search process. One scores the variously randomly generated coefficient vectors by the objective function, which does not have to be smooth and continuous with respect to the coefficient weights Ω. De Falco (1998) applied the genetic algorithm to nonlinear neural network estimation and found that his results "proved the effectiveness" of such algorithms for neural network estimation.

The main drawback of the genetic algorithm is that it is slow. For even a reasonable size or dimension of the coefficient vector Ω, the various combinations and permutations of elements of Ω that the genetic search may find optimal or close to optimal at various generations may become very large. This is another example of the well-known curse of dimensionality in nonlinear optimization. Thus, one needs to let the genetic algorithm run over a large number of generations — perhaps several hundred — to arrive at results that resemble unique and global minimum points.

Since the gradient-descent and simulated annealing methods rely on an arbitrary initialization of Ω, the best procedure for estimation may be a *hybrid approach*. One may run the genetic algorithm for a reasonable number of generations, say 100, and then use the final weight vector Ω as the initialization vector for the gradient-descent or simulated annealing minimization. One may repeat this process once more, with the final coefficient vector from the gradient-descent estimation entering a new population pool for selection, breeding, and mutation. Even this hybrid procedure is no sure thing, however.

Quagliarella and Vicini (1998) point out that hybridization may lead to better solutions than those obtainable using the two methods individually. These authors suggest the following alternative approaches:

1. The gradient-descent method is applied only to the best fit individual after many generations.

2. The gradient descent method is applied to several individuals, assigned by a selection operator.

3. The gradient descent method is applied to a number of individuals after the genetic algorithm has run many generations, but the selection is purely random.

Quagliarella and Vicini argue that it is not necessary to carry out the gradient-descent optimization until convergence, if one is going to repeat the process several times. The utility of the gradient-descent algorithm is its ability to improve the "individuals it treats" so "its beneficial effects can be obtained just performing a few iterations each time" [Quagliarella and Vicini (1998), p. 307].

The genetic algorithm and the hybridization method fit into a broader research agenda of evolutionary algorithms used not only for optimization but also for classification, or explaining the pattern or markets or organizations through time [see Bäck (1996)]. This is the estimation method used throughout this book. To level the playing field, we use this method not only for the neural network models but also for the competing models that require nonlinear estimation.

3.3   Repeated Estimation and Thick Models

The world of nonlinear estimation is a world full of traps, where we can get caught in local minimal or saddle points very easily. Thus, repeated estimation through hybrid genetic algorithm and gradient descent methods may be the safest check for the robustness of results after one estimation exercise with the hybrid approach.

For obtaining forecasts of particular variables, we must remember that neural network estimation, coupled with the genetic algorithm, even with the same network structure, never produces identical results, so that we should not put too much faith in particular point forecasts. Granger and Jeon (2002) have suggested "thick modeling" as a strategy for neural networks, particularly for forecasting. The idea is simple and straightforward. We should repeatedly estimate a given data set with a neural network. Since any neural network structure never gives identical results, we can use the same network specification, or we can change the specification of the network, or the scaling function, or even the estimation method, for different iterations on the network. What Granger and Jeon suggest is that we take a mean or trimmed mean of the forecasts of these alternative networks for our overall network forecast. They call this forecast a *thick model forecast*. We can also use this method for obtaining intervals for our forecasts of the network.

Granger and Jeon have pointed out an intriguing result from their studies of neural network performance, relative to linear models, for macroeconomic time series. They found that individual neural network models did not outperform simple linear models for most macro data, but thick models based on different neural networks uniformly outperformed the linear models for forecasting accuracy.

This approach is similar to bagging predictors in the broader artificial intelligence and machine learning literature [see Breiman (1996)]. With bagging, we can take a simple mean of various point forecasts coming from an ensemble of models. For classification, we take a plurality vote of the forecasts of multiple models. However, bagging is more extensive. The alternative forecasts may come not from different models per se, but from bootstrapping the initial training set. As we discuss in Section 4.2.8, bootstrapping involves resampling the original training set with replacement, and then taking repeated forecasts. Bagging is particularly useful if the data set exhibits instability or structural change. Combining the forecasts based on different randomly sampled subsets of the training set may give greater precision to the forecasting.

3.4   MATLAB Examples: Numerical Optimization and Network Performance

3.4.1   Numerical Optimization

To make these concepts about optimization more concrete and clear, we can take a simple problem, for which we can calculate an analytical solution. Assume we wish to optimize the following function with respect to inputs x and y:

$$z = .5x^2 + .5y^2 - 4x - 4y - 1 \qquad (3.30)$$

The solution can readily be obtained analytically, with $x^* = y^* = 4$, for the local minimum. A three-dimensional graph appears in Figure 3.2, with the solution for $x^* = y^* = 4$, illustrated by the arrow on the (x, y) grid.

A simple MATLAB program for calculating the global genetic algorithm search solution, the local simulated annealing solution, and the local quasi-Newton based on the BFGS algorithm appear, is given by the following sets of commands:

```
% Define simple function
z = inline('.5 * x(1) ^2 + .5 * x(2) ^2 - 4 * x(1) - 4 * x(2) - 1');
% Use random initialization
x0 = randn(1,2);
% Genetic algorithm parameters and execution-popsize, no. of
generations
maxgen = 100; popsize = 40;
xy_genetic = gen7f(z, x0, popsize,maxgen);
% Simulated annealing procedure (define temperature)
```

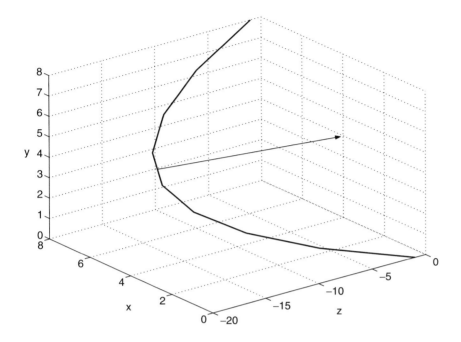

FIGURE 3.2. Sample optimization

```
TEMP = 500;
xy_simanneal = simanneal(z, xy_genetic, TEMP);
% BFGS Quasi-Newton Optimization Method
xy_bfgs = fminunc(z, xy_simanneal);
```

The solution for all three solution methods, the global genetic algorithm, the local search (using the initial conditions based on the genetic algorithm) and the quasi-Newton BFGS algorithm all yield results almost exactly equal to 4 for both x and y. While this should not be surprising, it is a useful exercise to check the accuracy of numerical methods by verifying how well they produce the true results obtained by analytical solution.

Of course, we use numerical methods precisely because we cannot obtain results analytically. Consider the following optimization problem, only slightly different from the previous function:

$$z = .5 \mid x \mid^{1.5} + .5 \mid x \mid^{2.5} + \cdots$$

$$.5 \mid y \mid^{1.5} + .5 \mid y \mid^{2.5} - 4x - 4y - 1 \tag{3.31}$$

Taking the partial derivatives with respect to x and y, we find the following first-order conditions:

$$.5 \cdot 1.5 \mid x \mid^{.5} + .5 \cdot \mid x \mid^{1.5} - 4 = 0 \tag{3.32}$$

$$.5 \cdot 1.5 \mid y \mid^{.5} + .5 \cdot \mid y \mid^{1.5} - 4 = 0$$

It should be clear that the optimal values x and y do not have closed-form or exact analytical solutions. The following MATLAB code solves this problem by the three algorithms:

```
% MATLAB Program for Minimization for Inline function z
z = inline('.5 * abs(x(1)) ^1.5 + .5 *abs(x(1)) ^2.5 + ...
.5 * abs(x(2)) ^1.5 + .5 * abs(x(2))^2.5 - 4 * x(1) - 4 * x(2) - 1');
% Initial guess of solution based on random numbers
x0 = randn(1,2);
% Initialization for Genetic Algorithm
maxgen = 100; popsize(50);
% Solution for genetic algorithm
xy_genetic = gen7f(z,x0, popsize, maxgen);
% Temperature for simulated annealing
TEMP = 500;
% Solution for simulated annealing
xy_simanneal = simanneal(z, xy_genetic, TEMP);
% BFGS Solution
xy_bfgs = fminunc(z, xy_simanneal);
```

Theoretically the solution values should be identical to each other. The results we obtain by the MATLAB process for the hybrid method of using the genetic algorithm, simulated annealing, and the quasi-Newton method, give values of $x = 1.7910746, y = 1.7910746$.

3.4.2 Approximation with Polynomials and Neural Networks

We can see how efficient neural networks are relative to linear and polynomial approximations with a very simple example. We first generate a standard normal random variable x of sample size 1000, and then generate a variable $y = [\sin(x)]^2 + e^{-x}$. We can then do a series of regressions with polynomial approximators and a simple neural network with two neurons, and compare the multiple correlation coefficients. We do this with the following set of MATLAB commands, which access the following functions for the orthogonal polynomials: *chedjudd.m, hermiejudd.m, legendrejudd.m,* and *laguerrejudd.m,* as well as the feedforward neural network program, *ffnet9.m.*

```
for j = 1:1000,
% Matlab Program For Assessing Approximation
randn('state',j);
x1 = randn(1000,1);
y 1= sin(x1).^2 + exp(-x1);
x = ((2 * x1) ./ (max(x1)-min(x1)))
    - ((max(x1)+min(x1))/(max(x1)-min(x1)));
y = ((2 * y1) ./ (max(y1)-min(y1)))
    - ((max(y1)+min(y1))/(max(y1)-min(y1)));
% Compute linear approximation
xols = [ones(1000,1) x];
bols = inv(xols'*xols)*xols'* y;
rsqols(j) = var(xols*bols)/var(y);
% Polynomial approximation
xp = [ones(1000,1) x x.^2];
bp = inv(xp'*xp)*xp'*y;
rsqp(j) = var(xp*bp)/var(y);
% Tchebeycheff approximation
xt = [ones(1000,1) chebjudd(x,3)];
bt = inv(xt'*xt)*xt'*y;
rsqt(j) = var(xt * bt)/var(y);
% Hermite approximation
xh = [ones(1000,1) hermitejudd(x,3)];
bh = inv(xh'*xh)*xh'*y;
rsqh(j)= var(xh * bh)/var(y);
% Legendre approximation
xl = [ones(1000,1) legendrejudd(x,3)];
bl = inv(xl'*xl)*xl'*y;
rsql(j)= var(xl * bl)/var(y);
% Leguerre approximation
xlg = [ones(1000,1) laguerrejudd(x,3)];
blg = inv(xlg'*xlg)*xlg'*y;
rsqlg(j)= var(xlg * blg)/var(y);
% Neural Network Approximation
data = [y x];
position = 1; % column number of dependent variable
architecture = [1 2 0 0]; % feedforward network with one hidden
    layer, with two neurons
geneticdummy = 1; % use genetic algorithm
maxgen =20; % number of generations for the genetic algorithm
percent = 1; % use 100 percent of data for all in-sample estimation
nlags = 0; % no lags for the variables
ndelay = 0; % no leads for the variables
niter = 20000; % number of iterations for quasi-Newton method
[sse, rsqnet01] = ffnet9(data, position, percent, nlags, ndelay,
    architecture, ...
```

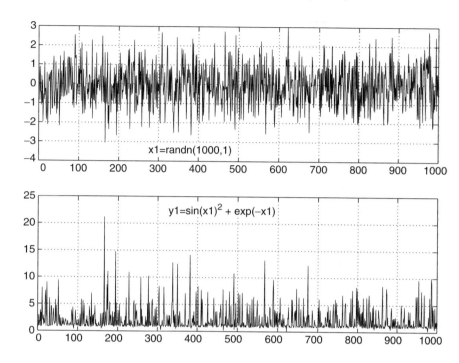

FIGURE 3.3. Sample nonlinear realization

```
geneticdummy, maxgen, niter) ;
rsqnet0(j) = rsqnet01(2);
RSQ(j,:) = [rsqols(j) rsqp(j) rsqt(j) rsqh(j) rsql(j) rsqlg(j)
    rsqnet0(j)]
end
```

One realization of the variables $[y\ x]$ appears in Figure 3.3. While the process for the variable x is a standard random realization, we see that the process for y contains periodic jumps as well as periods of high volatility followed by low volatility. Such properties are common in financial markets, particularly in emerging market countries.

Table 3.3 gives the results for the goodness of fit or R^2 statistics for this base set of realizations, as well as the mean and standard deviations of this measure for 1000 additional draws of the same sample length. We compare second-order polynomials with a simple network with two neurons. This table brings home several important results. First, there are definite improvements in abandoning pure linear approximation. Second, the power polynomial and the orthogonal polynomials give the same results. There is no basis for preferring one over the other. Third, the neural network, a

TABLE 3.3. Goodness of Fit Tests of Approximation Methods

Approximation	R^2: Base Run	Mean R^2 – 1000 Draws (std. deviation)
Linear	.49	.55 (.04)
Polynomial-Order 2	.85	.91 (.03)
Tchebycheff Polynomial-Order 2	.85	.91 (.03)
Hermite-Order 2	.85	.91 (.03)
Legendre-Order 2	.85	.91 (.03)
Laguerre-Order 2	.85	.91 (.03)
Neural Network: FF, 2 neurons, 1 layer	.99	.99 (.005)

very simple neural network, is superior to the polynomial expansions, and delivers a virtually perfect fit. Finally, the neural network is much more precise, relative to the other methods, across a wide set of realizations.

3.5 Conclusion

This chapter shows how the introduction of nonlinearity makes the estimation problem much more challenging and time-consuming than the case of the standard linear model. But it also makes the estimation process much more interesting. Given that we can converge to many different results or parameter values, we have to find ways to differentiate the good from the bad, or the better from a relatively worse set of estimates. Engineers have been working with nonlinear optimization for many decades, and this chapter shows how we can apply many of the existing evolutionary global or hybrid search methods for neural network estimation. We need not resign ourselves to the high risk of falling into locally optimal results.

3.5.1 MATLAB Program Notes

Optimization software is quite common. The MATLAB function *fminunc.m*, for unconstrained minimization, part of the Optimization Toolbox, is the one used for the quasi-Newton gradient-based methods. It has lots of options, such as the specification of tolerance criteria and the maximum number of iterations. This function, like most software, is a minimization function. For maximizing a likelihood function, we minimize the negative of the likelihood function.

The genetic algorithm used above is *gen7f.m*. The function requires four inputs, including the name of the function being minimized. The function being optimized, in turn, must have as its first output the criterion to be minimized, such as a sum of squared errors, or the negative of the likelihood function.

The function *simanneal.m* requires the specification of the function, coefficient matrix, and initial temperature. Finally, the orthogonal polynomial operators, *chedjudd.m*, *hermiejudd.m*, *legendrejudd.m*, and *laguerrejudd.m* are also available.

The scaling functions for transforming variables to ranges between [0,1] or [−1,1] are in the MATLAB Neural Net Toolbox, *premnmx.m*.

The scaling function for the transformation suggested by Helge Petersohn is given by *hsquasher.m*. The reverse transformation is given by *helgeyx.m*.

3.5.2 Suggested Exercises

As a follow-up to the exercises on minimization, we can do more comparisons of the accuracy of the simulated annealing and genetic algorithm with benchmark true analytical solutions for a variety of functions. Simply use the MATLAB Symbolic Toolbox *funtool.m* to find the true minimum for a host of functions by setting the first derivative to zero. Then use *simanneal.m* and *gen7f.m* to find the numerical approximate solutions.

4
Evaluation of Network Estimation

So far we have discussed the structure or architecture of a network, as well as the ways of training or estimating the coefficients or weights of a network. How do we interpret the results obtained from these networks, relative to what we can obtain from a linear approximation?

There are three sets of criteria: in-sample criteria, out-of-sample criteria, and common sense based on tests of significance and the plausibility of the results.

4.1 In-Sample Criteria

When evaluating the regression, we first want to know how well a model fits the actual data used to obtain the estimates of the coefficients. In the neural network literature, this is known as supervised training. We supervise the network, insofar as we evaluate it by how well it fits actual data.

The first overall statistic is a measure of goodness of fit. The Hannan-Quinn information is a method for handicapping this measure for competing models that have different numbers of parameters.

The other statistics relate to properties of the regression residuals. If the model is indeed a good fit, and thus well specified, then there should be nothing further to learn from the residuals. The residuals should simply represent "white noise," or uncorrelated meaningless information — like listening to a fan or air conditioner, which we readily and easily ignore.

4.1.1   Goodness of Fit Measure

The most commonly used measure of overall goodness of fit of a model is the multiple correlation coefficient, also known as the R-squared coefficient. It is simply the ratio of the variance of the output predicted by the model relative to the true or observed output:

$$R^2 = \frac{\sum_{t=1}^{T}(\widehat{y}_t - \overline{y}_t)^2}{\sum_{t=1}^{T}(y_t - \overline{y}_t)^2} \tag{4.1}$$

This value falls in the interval $[0, 1]$ if there is a constant term in the model.

4.1.2   Hannan-Quinn Information Criterion

Of course, we can generate progressively higher values of the R^2 statistic by using a model with an increasingly larger number of parameters. One way to modify the R^2 statistic is to make use of the Hannan-Quinn (1979) information criterion, which handicaps or "punishes" the performance of a model for the number of parameters, k, it uses:

$$hqif = \left[\ln\left(\sum_{t=1}^{T}\frac{(y_t - \widehat{y}_t)^2}{T}\right)\right] + \frac{k\{\ln[\ln(T)]\}}{T} \tag{4.2}$$

The criterion is simply to choose the model with the lowest value. Note that the hqif statistic punishes a given model by a factor of $k\{\ln[\ln(T)]\}/T$, the logarithm of the logarithm of the number of observations, T, multiplied by the number of parameters, k, divided by T. The Akaike criterion replaces the second term on the right-hand side of equation (4.2) with the variable $2k/T$, whereas the Schwartz criterion replaces the same term with the value $k[\ln(T)]/T$. We work with the Hannan-Quinn statistic rather than the Akaike or Schwartz criteria, on the grounds that *virtu stat in media*. The Hannan-Quinn statistic usually punishes a model with more parameters more than the Akaike (1974) statistic, but not as severely as the Schwartz (1978) statistic.[1]

4.1.3   Serial Independence: Ljung-Box and McLeod-Li Tests

If a model is well specified, the residuals should have no systematic pattern in their first or second moments. Tests for serial independence and constancy of variance, or homoskedasticity, are the first steps for evaluating whether or not there is any meaningful information content in the residuals.

---

[1]The penalty factor attached to the number of parameters in a model is known as the *regularization term* and represents a control or check over the effective complexity of a model.

The most commonly used statistic tests for serial independence against the alternative hypothesis of first-order autocorrelation is the well-known, but elementary, Durbin-Watson (DW) test.

$$DW = \frac{\sum_{t=2}^{T} [\hat{\varepsilon}_t - \hat{\varepsilon}_{t-1}]^2}{\sum_{t=1}^{T} \hat{\varepsilon}_t^2} \tag{4.3}$$

$$\approx 2 - 2\rho_1(\hat{\varepsilon}) \tag{4.4}$$

where $\rho_1(\hat{\varepsilon})$ represents the first order autocorrelation coefficient.

In the absence of autocorrelation, each residual represents a surprise which is unpredictable from the past data. The autocorrelation function is given by the following formula, for different lag lengths m:

$$\rho_m(\hat{\varepsilon}) = \frac{\sum_{t=m+1}^{T} \hat{\varepsilon}_t \hat{\varepsilon}_{t-m}}{\sum_{t=1}^{T} \hat{\varepsilon}_t^2} \tag{4.5}$$

Ljung and Box (1978) put forward the following test statistic, known as the Ljung-Box Q-statistic, for examining the joint significance of the first M residual autocorrelations, with an asymptotic Chi-squared distribution having M degrees of freedom:

$$Q(M) = (T)(T+2) \sum_{m \equiv 1}^{M} \frac{\rho_m^2(\hat{\varepsilon})}{(T-m)} \tag{4.6}$$

$$\sim \chi^2(M) \tag{4.7}$$

If a model does not pass the Ljung-Box Q test, there is usually a need for correction. We can proceed in two ways. One is simply to add more lags of the dependent variable as regressors or input variables. In many cases, this takes care of serial dependence. An alternative is to respecify the error structure itself as a moving average (MA process). In dynamic models, in which we forecast the inflation rate over several quarters, we build in by design a moving average process into the disturbance or innovation terms. In this case, the inflation we forecast in January is the inflation rate from next January to this January. In the next quarter, we forecast the inflation rate from next April to this April. However, the forecast from next April to this April will depend a great deal on the forecast error from next January to this past January. Yet in forecasting exercises, often we are most interested in forecasting over several periods rather than for one period into the future, so purging the estimates of serial dependence is extremely important before we do any assessment of the results. This is especially true when we compare a linear model with the neural network alternative. The linear model first should be purged of serial dependence either by the use of a liberal lag structure or by an MA specification for the error term, before we can make any meaningful comparison with alternative functional forms.

Adding more lags of the dependent variable is easy enough. The use of the MA specification requires the following transformation of the error term for a linear model with an MA component of order p:

$$y_t = \sum_{k=0}^{K} \beta_k x_{k,t} + \epsilon_t \qquad (4.8a)$$

$$\epsilon_t = \eta_t - \rho_1 \eta_{t-1} - \ldots \rho_p \eta_{t-p} \qquad (4.8b)$$

$$\eta_t \tilde{\ } N(0, \sigma^2) \qquad (4.8c)$$

Joint estimation of the coefficient set $\{\beta_k\}$ and $\{\rho_i\}$ is done by maximum likelihood estimation. Tsay (2002, p. 46) distinguishes between conditional and exact likelihood estimation of the MA terms $\{\rho_i\}$. With conditional estimation, for the first periods, $\{t = 1, \ldots, t*)$, with $t* \leq p$, we simply assume that the error terms are zero. Exact estimation takes a more careful approach. For period $t = 1$, the shocks $\eta_{t-i}, i = 1, \ldots, p$, are set at zero. However, for $t = 2$, η_{t-1} is known, so the realized error is used, while the other shocks $\eta_{t-i}, i = 2, \ldots, p$, are set at zero. We follow a similar process for the observations for $t \leq p$. For $t > p$, of course, we can use the realized values of the errors. In many cases, as Tsay points out, the differences in the coefficient values and resulting Q statistics from conditional and exact likelihood estimation is very small.

Since the squared residuals of the model are used to compute standard errors of estimates of the model, one can apply an extension of the Ljung-Box Q statistic to test for homoskedasticity, or constancy of the variance, of the residuals against an unspecified alternative. In a well-specified model, the variance should be constant. This test is the McLeod and Li (1983), and tests for autocorrelation of the squared residuals, with the same distribution and degrees of freedom as the Q statistic.

$$McL(M) = (T)(T + 2) \sum_{m=1}^{M} \frac{\rho_m^2(\hat{\epsilon}^2)}{(T - m)} \qquad (4.9)$$

$$\sim \chi^2(M) \qquad (4.10)$$

In many cases, we will find that correcting for the serial dependence in the levels of the residuals is also a correction for serial dependence in the squared residuals. Alternatively, a linear model may show a significant Q

TABLE 4.1. Engle-Ng Test of Symmetry of Residuals

Definition	Operation
Standardized errors	$\widetilde{\epsilon}_\tau = \widehat{\epsilon}_\tau / \sigma_{\widehat{\epsilon}_\tau}$
Squared standardized errors	$\{\widetilde{\epsilon}_\tau^2\}$
Positive indicators	$\widetilde{\epsilon}_\tau^+ = 1$ if $\widetilde{\epsilon}_\tau > 0, 0$ otherwise
Negative indicators	$\widetilde{\epsilon}_\tau^- = 1$ if $\widetilde{\epsilon}_\tau < 0, 0$ otherwise
Positive valued errors	$\eta_\tau^+ = \widetilde{\epsilon}_\tau \cdot \widetilde{\epsilon}_\tau^+$
Negative valued errors	$\eta_\tau^- = \widetilde{\epsilon}_\tau \cdot \widetilde{\epsilon}_\tau^-$
Regression	$y_\tau = \widetilde{\epsilon}_\tau^2, x_\tau = [1 \ \widetilde{\epsilon}_\tau^- \ \eta_\tau^+ \ \eta_\tau^-]$
Engle-Ng LM statistic	$LM = (T-1) \cdot R^2$
Distribution	$LM \sim \chi^2(3)$

statistic for the McLeod-Li test whereas a neural network alternative may not. The point is that for making a fair comparison between a linear and network model, the most important issue is to correct the linear model for serial dependence in the raw, rather than the squared, value of the residuals.

4.1.4 Symmetry

In addition to serial independence and constancy of variance, symmetry of the residuals is also a desired property if they indeed represent purely random shocks. Symmetry is an important issue, of course, if the model is going to be used for simulation with symmetric random shocks. However, violation of the symmetry assumption is not as serious as violation of serial independence or constancy of variance.

The test for symmetry of residuals proposed by Engle and Ng (1993) is shown in Table 4.1.

4.1.5 Normality

Besides having properties of serial independence, constancy of variance, and symmetry, residuals are usually assumed to come from a Gaussian or normal distribution. One well-known test, the Jarque-Bera statistic, starts from the assumption that a normal distribution has zero skewness and a kurtosis of 3.

Given the residual vector $\widehat{\epsilon}$, the Jarque-Bera (1980) statistic is given by the following formula and distribution:

$$JB(\widehat{\epsilon}) = \frac{T-k}{6} \left\{ SK(\widehat{\epsilon})^2 + .25(KR(\widehat{\epsilon}) - 3)^2 \right\} \qquad (4.11)$$

$$\sim \chi^2(2)$$

where $SK(\widehat{\epsilon})$ and $KR(\widehat{\epsilon})$, for skewness and kurtosis, are defined as follows:

$$SK(\widehat{\epsilon}) = \frac{1}{T} \sum_{t=1}^{T} \left(\frac{\widehat{\epsilon}_i - \overline{\widehat{\epsilon}}}{\sigma_{\widehat{\epsilon}}} \right)^3 \qquad (4.12)$$

$$KR(\widehat{\epsilon}) = \frac{1}{T} \sum_{t=1}^{T} \left(\frac{\widehat{\epsilon}_i - \overline{\widehat{\epsilon}}}{\sigma_{\widehat{\epsilon}}} \right)^4 \qquad (4.13)$$

while $\overline{\widehat{\epsilon}}$ and $\sigma_{\widehat{\epsilon}}$ represent the estimated mean and standard deviation of the residual vector $\widehat{\epsilon}$.

How important is the normality assumption, or how serious is a violation of the normality assumption? The answer depends on the purpose of the estimation. If the estimated model is going to be used for simulating models subject to random normal disturbances, then it would be good to have normal randomly distributed residuals in the estimated model.

4.1.6   Neural Network Test for Neglected Nonlinearity: Lee-White-Granger Test

Lee, White, and Granger (1992) proposed the use of artificially generated neural networks for testing for the presence of neglected nonlinearity in the regression residuals of any estimated model. The test works with the regressions residuals and the inputs of the model, and seeks to find out if any of the residuals can be explained by nonlinear transformations of the input variables. If they can be explained, there is neglected nonlinearity.

Since the precise form of the nonlinearity is unspecified, Lee, White, and Granger propose a neural network approach, but they leave aside the time-consuming estimation process for the neural network. Instead, the coefficients or weights linking the inputs to the neurons are generated randomly.

The Lee, White, Granger (L-W-G) test is rather straightforward, and proceeds in six steps:

1. From the initial model, obtain the residuals and the input variables.

2. Generate a set of neuron regressors from the inputs, with randomly generated weights for the input variables.

3. Regress the residuals on the neurons, and obtain the multiple correlation coefficients.

4. Repeat this process 1000 times.

TABLE 4.2. Lee-White-Granger Test of Neglected Nonlinearity

Definition	Operation
Obtain residuals and inputs	e, x
Randomly generate P sets of coefficients for x	β_i
Generate P neurons $n_1, n_2, \ldots, n_p$	$n_p = \frac{1}{1+e^{-\beta_p x}}$
Regress e on the P neurons	$e = b_1 n_1 +, \ldots, b_p n_p$
Obtain multiple correlation coefficient	R_1^2
Repeat process 1000 times	$R_1^2, R_2^2, \ldots, R_{1000}^2$
Assess significance of coefficients	$F(R_1^2), \ldots, F(R_{1000}^2)$
Count significant F statistics	$I_i = 1 \Leftrightarrow F(R_1^2) > F^*$
Decision: Reject if more than 5% significant	

5. Assess the significance of the multiple correlation coefficients by F statistics.

6. If these coefficients are significant more than 5% of the time, there is a case for neglected nonlinearity.

For convenience, these steps are summarized in Table 4.2.

This test is similar to the White (1980) test for heteroskedasticity. This test is a regression of the squared residuals on a polynomial expansion of the regressors or input variables. In the White test, we specify the power of the polynomial, with the option to include or exclude the cross terms in the polynomial expansion of the input variables.

The intuition behind the L-W-G test is that if there is any neglected nonlinearity in the residuals, some combination of neural network transformations of the inputs should be able to explain or detect it by approximating it well, since neural networks are adept at approximating unknown nonlinear functions. Since linear regressions of the residuals are done on the randomly generated neurons, the test proceeds very rapidly. If, after a large number of repeated trials with randomly generated neurons, no significant relations between the neurons and the residuals emerge, one can be confident that there are no neglected nonlinearities.

4.1.7 Brock-Deckert-Scheinkman Test for Nonlinear Patterns

Brock, Deckert, and Scheinkman (1987), further elaborated in Brock, Deckert, Scheinkman, and LeBaron (1996), propose a test for detecting nonlinear patterns in time series. Following Kocenda (2001), the null hypothesis is that the data are independently and identically distributed

TABLE 4.3. BDS Test of IID Process

Definition	Operation
Form m-dimensional vector, x_t^m	$x_t^m = x_t, \ldots, x_{t+m}, t = 1, \ldots, T_{m-1}, T_{m-1} = T - m$
Form m-dimensional vector, x_s^m	$x_s^m = x_s, \ldots, x_{s+m}, s = t + 1, \ldots, T_m, T_m = T - m + 1$
Form indicator function	$I_\varepsilon(x_t^m, x_s^m) = \max\limits_{i=0,1,\ldots,m-1} \mid x_{t+1} - x_{s+i} \mid < \varepsilon$
Calculate correlation integral	$C_{m,T}(\varepsilon) = 2 \sum_{t=1}^{T_{m-1}} \sum_{s=t+1}^{T_m} \frac{I_\varepsilon(x_t^m, x_s^m)}{T_m(T_{m-1}-1)}$
Calculate correlation integral	$C_{1,T}(\varepsilon) = 2 \sum_{t=1}^{T-1} \sum_{s=t+1}^{T} \frac{I_\varepsilon(x_t^1, x_s^1)}{T(T-1)}$
Form Numerator	$\sqrt{T} \left[C_{m,T}(\varepsilon) - C_{1,T}(\varepsilon)^m \right]$
Sample Standard Dev. of Numerator	$\sigma_{m,T}(\varepsilon)$
Form BDS Statistic	$BDS_{m,T}(\varepsilon) = \frac{\sqrt{T}\left[C_{m,T}(\varepsilon) - C_{1,T}(\varepsilon)^m\right]}{\sigma_{m,T}(\varepsilon)}$
Distribution	$BDS_{m,T}(\varepsilon) \sim N(0,1)$

(iid) processes. This test, known as the BDS test, is unique in its ability to detect nonlinearities independently of linear dependencies in the data.

The test rests on the correlation integral, developed to distinguish between chaotic deterministic systems and stochastic systems. The procedure consists of taking a series of m-dimensional vectors from a time series, at time $t = 1, 2, \ldots, T - m$, where T is the length of the time series. Beginning at time $t = 1$ and $s = t + 1$, the pairs (x_t^m, x_s^m) are evaluated by an indicator function to see if their maximum distance, over the horizon m, is less than a specified value ε. The correlation integral measures the fraction of pairs that lie within the tolerance distance for the embedding dimension m.

The BDS statistic tests the difference between the correlation integral for embedding dimension m, and the integral for embedding dimension 1, raised to the power m. Under the null hypothesis of an iid process, the BDS statistic is distributed as a standard normal variate.

Table 4.3 summarizes the steps for the BDS test.

Kocenda (2002) points out that the BDS statistic suffers from one major drawback: the embedding parameter m and the proximity parameter ε must be chosen arbitrarily. However, Hsieh and LeBaron (1988a, b, c) recommend choosing ε to be between .5 and 1.5 standard deviations of the data. The choice of m depends on the lag we wish to examine for serial dependence. With monthly data, for example, a likely candidate for m would be 12.

4.1.8   Summary of In-Sample Criteria

The quest for a high measure of goodness of fit with a small number of parameters with regression residuals that represent random white noise is a difficult challenge. All of these statistics represent tests of specification error, in the sense that the presence of meaningful information in the residuals indicates that key variables are omitted, or that the underlying true functional form is not well approximated by the functional form of the model.

4.1.9   MATLAB Example

To give the preceding regression diagnostics clearer focus, the following MATLAB code randomly generates a time series $y = \sin(x)^2 + \exp(-x)$ as a nonlinear function of a random variable x, then uses a linear regression model to approximate the model, and computes the in-sample diagnostic statistics. This program makes use of functions *ols1.m*, *wnnest1.m*, and *bds.m*, available on the webpage of the author.

```
% Create random regressors, constant term,
% and dependent variable
for i = 1:1000,
randn('state',i);
xxx = randn(1000,1);
x1 = ones(1000,1);
x = [x1 xxx];
y = sin(xxx).^2 + exp(-xxx);
% Compute ols coefficients and diagnostics
[beta, tstat, rsq, dw, jbstat, engle, ...
lbox, mcli] = ols1(x,y);
% Obtain residuals
residuals = y - x * beta;
sse = sum(residuals .^2);
nn = length(residuals);
kk = length(beta);
% Hannan-Quinn Information Criterion
k = 2;
hqif = log(sse/nn) + k * log(log(nn))/nn;
% Set up Lee-White-Granger test
neurons = 5;
nruns = 1000;
% Nonlinearity Test
[nntest, nnsum] = wnntest1(residuals, x, neurons, nruns);
% BDS Nonlinearity Test
[W, SIG] = bds1(residuals);
RSQ(i) = rsq;
DW(i) = dw;
```

TABLE 4.4. Specification Tests

Test Statistic	Mean	% of Significant Tests
JB-Marginal significance	0	100
EN-Marginal significance	.56	3.7
LB-Marginal significance	.51	4.5
McL-Marginal Significance	.77	2.1
LWG-No. of Significant Regressions	999	99
BDS-Marginal Significance	.47	6.6

```
JBSIG(i) = jbstat(2);
ENGLE(i) = engle(2);
LBOX(i) = lbox(2);
MCLI(i) = mcli(2);
NNSUM(i) = nnsum;
BDSSIG(i) = SIG;
HQIF(i) = hqif;
SSE(i) = sse;
end
```

The model is nonlinear, and estimation with linear least squares clearly is a misspecification. Since the diagnostic tests are essentially various types of tests for specification error, we examine in Table 4.4 which tests pick up the specification error in this example. We generate data series of sample length 1000 for 1000 different realizations or experiments, estimate the model, and conduct the specification tests.

Table 4.4 shows that the JB and the LWG are the most reliable for detecting misspecification for this example. The others do not do nearly as well: the BDS tests for nonlinearity are significant 6.6% of the time, and the LB, McL, and EN tests are not even significant for 5% of the total experiments. In fairness, the LB and McL tests are aimed at serial correlation, which is not a problem for these simulations, so we would not expect these tests to be significant. Table 4.4 does show, very starkly, that the Lee-White-Granger test, making use of neural network regressions to detect the presence of neglected nonlinearity in the regression residuals, is highly accurate. The Lee-White-Granger test picks up neglected nonlinearity in 99% of the realizations or experiments, while the BDS test does so in 6.6% of the experiments.

4.2   Out-of-Sample Criteria

The real acid test for the performance of alternative models is its out-of-sample forecasting performance. Out-of-sample tests evaluate how well competing models generalize outside of the data set used for estimation. Good in-sample performance, judged by the R^2 or the Hannan-Quinn statistics, may simply mean that a model is picking up peculiar or idiosyncratic aspects of a particular sample or over-fitting the sample, but the model may not fit the wider population very well.

To evaluate the out-of-sample performance of a model, we begin by dividing the data into an in-sample estimation or training set for obtaining the coefficients, and an out-of-sample or test set. With the latter set of data, we plug in the coefficients obtained from the training set to see how well they perform with the new data set, which had no role in calculating of the coefficient estimates.

In most studies with neural networks, a relatively high percentage of the data, 25% or more, is set aside or withheld from the estimation for use in the test set. For cross-section studies with large numbers of observations, withholding 25% of the data is reasonable. In time-series forecasting, however, the main interest is in forecasting horizons of several quarters or one to two years at the maximum. It is not usually necessary to withhold such a large proportion of the data from the estimation set.

For time-series forecasting, the out-of-sample performance can be calculated in two ways. One is simply to withhold a given percentage of the data for the test, usually the last two years of observations. We estimate the parameters with the training set, use the estimated coefficients with the withheld data, and calculate the set of prediction errors coming from the withheld data. The errors come from one set of coefficients, based on the fixed training set and one fixed test set of several observations.

4.2.1 Recursive Methodology

An alternative to a once-and-for-all division of the data into training and test sets is the recursive methodology, which Stock (2000) describes as a series of "simulated real time forecasting experiments." It is also known as estimation with a "moving" or "sliding" window. In this case, period-by-period forecasts of variable y at horizon h, $\widehat{y}_{t+h}$, are conditional only on data up to time t. Thus, with a given data set, we may use the first half of the data, based on observations $\{1, \ldots, t^*\}$ for the initial estimation, and obtain an initial forecast $\widehat{y}_{t^*+h}$. Then we re-estimate the model based on observations $\{1, \ldots, t^* + 1\}$, and obtain a second forecast error, $\widehat{y}_{t^*+1+h}$. The process continues until the sample is covered. Needless to say, as Stock (2000) points out, the many re-estimations of the model required by this approach can be computationally demanding for nonlinear models. We call this type of recursive estimation an *expanding window*. The sample size, of course, becomes larger as we move forward in time.

An alternative to the expanding window is the moving window. In this case, for the first forecast we estimate with data observations $\{1, \ldots, t^*\}$, and obtain the forecast $\widehat{y}_{t^*+h}$ at horizon h. We then incorporate the observation at t^*+1, and re-estimate the coefficients with data observations $\{2,\ldots,t^*+1\}$, and not $\{1,\ldots,t^*+1\}$. The advantage of the moving window is that as data become more distant in the past, we assume that they have little or no predictive relevance, so they are removed from the sample.

The recursive methodology, as opposed to the once-and-for-all split of the sample, is clearly biased toward a linear model, since there is only one forecast error for each training set. The linear regression coefficients adjust to and approximate, step-by-step in a recursive manner, the underlying changes in the slope of the model, as they forecast only one step ahead. A nonlinear neural network model, in this case, is challenged to perform much better. The appeal of the recursive linear estimation approach is that it reflects how econometricians do in fact operate. The coefficients of linear models are always being updated as new information becomes available, if for no other reason, than that linear estimates are very easy to obtain. It is hard to conceive of any organization using information a few years old to estimate coefficients for making decisions in the present. For this reason, evaluating the relative performance of neural nets against recursively estimated linear models is perhaps the more realistic match-up.

4.2.2 Root Mean Squared Error Statistic

The most commonly used statistic for evaluating out-of-sample fit is the root mean squared error (rmsq) statistic:

$$rmsq = \sqrt{\frac{\sum_{\tau=1}^{\tau^*}(y_\tau - \widehat{y}_\tau)^2}{\tau^*}} \tag{4.14}$$

where τ^* is the number of observations in the test set and $\{\widehat{y}_\tau\}$ are the predicted values of $\{y_\tau\}$. The out-of-sample predictions are calculated by using the input variables in the test set $\{x_\tau\}$ with the parameters estimated with the in-sample data.

4.2.3 Diebold-Mariano Test for Out-of-Sample Errors

We should select the model with the lowest root mean squared error statistic. However, how can we determine if the out-of-sample fit of one model is significantly better or worse than the out-of-sample fit of another model? One simple approach is to keep track of the out-of-sample points in which model A beats model B.

A more detailed solution to this problem comes from the work of Diebold and Mariano (1995). The procedure appears in Table 4.5.

TABLE 4.5. Diebold-Mariano Procedure

Definition	Operation
Errors	$\{\widehat{\epsilon}_\tau\}, \{\widehat{\eta}_\tau\}$
Absolute differences	$z_\tau = |\widehat{\eta}_\tau| - |\widehat{\epsilon}_\tau|$
Mean	$\overline{z} = \frac{\sum_{\tau=1}^{\tau*} z_\tau}{\tau*}$
Covariogram	$c = [Cov(z_\tau, z_{\tau-p,}), Cov(z_\tau, z_\tau,), Cov(z_\tau, z_{\tau+p,})]$
Mean	$\overline{c} = \sum c/(p+1)$
DM statistic	$DM = \frac{\overline{z}}{\overline{c}} \sim N(0,1), H_0 : E(z_\tau) = 0$

As shown above, we first obtain the out-of-sample prediction errors of the benchmark model, given by $\{\epsilon_\tau\}$, as well as those of the competing model, $\{\eta_\tau\}$.

Next, we compute the absolute values of these prediction errors, as well as the mean of the differences of these absolute values, z_τ. We then compute the covariogram for lag/lead length p, for the vector of the differences of the absolute values of the predictive errors. The parameter $p < \tau^*$ is the length of the out-of-sample prediction errors.

In the final step, we form a ratio of the means of the differences over the covariogram. The DM statistic is distributed as a standard normal distribution under the null hypothesis of no significant differences in the predictive accuracy of the two models. Thus, if the competing model's predictive errors are significantly lower than those of the benchmark model, the DM statistic should be below the critical value of -1.69 at the 5% critical level.

4.2.4 Harvey, Leybourne, and Newbold Size Correction of Diebold-Mariano Test

Harvey, Leybourne, and Newbold (1997) suggest a size correction to the DM statistic, which also allows "fat tails" in the distribution of the forecast errors. We call this modified Diebold-Mariano statistic the MDM statistic. It is obtained by multiplying the DM statistic by the correction factor CF, and it is asymptotically distributed as a Student's t with $\tau^* - 1$ degrees of freedom. The following equation system summarizes the calculation of the MDM test, with the parameter p representing the lag/lead length of the covariogram, and τ^* the length of the out-of-sample forecast set:

$$CF = \frac{\tau^* + 1 - 2p + p(1-p)/\tau^*}{\tau^*} \tag{4.15}$$

$$MDM = CF \cdot DM \sim t_{\tau^*-1}(0,1) \tag{4.16}$$

4.2.5   Out-of-Sample Comparison with Nested Models

Clark and McCracken (2001), Corradi and Swanson (2002), and Clark and West (2004) have proposed tests for comparing out-of-sample accuracy for two models, when the competing models are nested. Such a test is especially relevant if we wish to compare a feedforward network with jump connections (containing linear as well as logsigmoid neurons) with a simple restricted linear alternative, given by the following equations:

$$\text{Restricted Model: } y_t = \sum_{k=1}^{K} \alpha_k x_{k,t} + \epsilon_t \tag{4.17}$$

$$\text{Alternative Model: } y_t = \sum_{k=1}^{K} \beta_k x_{k,t} + \sum_{j=1}^{J} \gamma_j N_{j,t} + \eta_t \tag{4.18}$$

$$N_{j,t} = \frac{1}{1 + \exp[-(\sum_{k=1}^{K} \delta_{j,k} x_{k,t})]} \tag{4.19}$$

where the first restricted equation is simply a linear function of K parameters, while the second unrestricted network is a nonlinear function with $K + JK$ parameters. Under the null hypothesis of equal predictive ability of the two models, the difference between the squared prediction errors should be zero. However, Todd and West point out that under the null hypothesis, the mean squared prediction error of the null model will often or likely be smaller than that of the alternative model [Clark and West (2004), p. 6]. The reason is that the mean squared error of the alternative model will be pushed up by noise terms reflecting "spurious small sample fit" [Clark and West (2004), p. 8]. The larger the number of parameters in the alternative model, the larger the difference will be.

Clark and West suggest a procedure for correcting the bias in out-of-sample tests. Their paper does not have estimated parameters for the restricted or null model — they compare a more extensive model against a simple random walk model for the exchange rate. However, their procedure can be used for comparing a pure linear restricted model against a combined linear and nonlinear alternative model as above. The procedure is a correction to the mean squared prediction error of the unrestricted model by an adjustment factor ψ_{ADJ}, defined in the following way, for the case of the neural network model.

The mean squared prediction errors of the two models are given by the following equations, for forecasts $\tau = 1, \ldots, T^*$:

$$\sigma_{RES}^2 = (T^*)^{-1} \sum_{\tau=1}^{T^*} \left[y_\tau - \sum_{k=1}^{K} \widehat{\beta}_k x_{k,\tau} \right]^2 \tag{4.20}$$

$$\sigma^2_{NET} = (T^*)^{-1} \sum_{\tau=1}^{T^*} \left[y_\tau - \sum_{k=1}^{K} \widehat{\alpha}_k x_{k,\tau} - \sum_{j=1}^{J} \widehat{\gamma}_j \left(\frac{1}{1+\exp[-(\sum_{k=1}^{K} \widehat{\delta}_{j,k} x_{k,\tau})]} \right) \right]^2$$

(4.21)

The null hypothesis of equal predictive performance is obtained by comparing σ^2_{NET} with the following adjusted mean squared error statistic:

$$\sigma^2_{ADJ} = \sigma^2_{NET} - \psi_{ADJ}$$

(4.22)

The test statistic under the null hypothesis of equal predictive performance is given by the following expression:

$$\widehat{f} = \sigma^2_{RES} - \sigma^2_{ADJ}$$

(4.23)

The approximate distribution of this statistic, multiplied by the square root of the size of the out-of-sample set, is given by normal distribution with mean 0 and variance V:

$$(T^*)^{.5} \widehat{f} \, \tilde{} \, \phi(\mathbf{0}, \mathbf{V})$$

(4.24)

The variance is computed in the following way:

$$V = 4 \cdot (T^*)^{-1} \sum_{\tau=1}^{T^*} \left[\left(y_\tau - \sum_{k=1}^{K} \widehat{\beta}_k x_{k,\tau} \right) \left(\sum_{j=1}^{J} \widehat{\gamma}_j N_{j,\tau} \right) \right]^2$$

(4.25)

Clark and West point out that this test is one-sided: if the restrictions of the linear model were not true, the forecasts from the network model would be superior to those of the linear model.

4.2.6   Success Ratio for Sign Predictions: Directional Accuracy

Out-of-sample forecasts can also be evaluated by comparing the signs of the out-of-sample predictions with the true sample. In financial time series, this is particularly important if one is more concerned about the sign of stock return predictions rather than the exact value of the returns. After all, if the out-of-sample forecasts are correct and positive, this would be a signal to buy, and if they are negative, a signal to sell. Thus, the correct sign forecast reflects the market timing ability of the forecasting model.

Pesaran and Timmermann (1992) developed the following test of directional accuracy (DA) for out-of-sample predictions, given in Table 4.6.

TABLE 4.6. Pesaran-Timmerman Directional Accuracy (DA) Test

Definition	Operation
Calculate out of sample predictions, m periods	$\widehat{y}_{n+j}, j = 1, \ldots, m$
Compute indicator for correct sign	$I_j = 1$ if $\widehat{y}_{n+j} \cdot y_{n+j} > 0, 0$ otherwise
Compute success ratio (SR)	$SR = \frac{1}{m} \sum_{j=1}^{m} I_j$
Compute indicator for true values	$I_j^{true} = 1$ if $y_{n+j} > 0, 0$ otherwise
Compute indicator for predicted values	$I_j^{pred} = 1$ if $\widehat{y}_{n+j} > 0, 0$ otherwise
Compute means $P, \widehat{P}$	$P = \frac{1}{m} \sum_{j=1}^{m} I_j^{true}, \widehat{P} = \frac{1}{m} \sum_{j=1}^{m} I_j^{pred}$
Compute success ratio under independence (SRI)	$SRI = P \cdot \widehat{P} - (1 - P) \cdot (1 - \widehat{P})$
Compute variance for SRI	$var(SRI) = \frac{1}{m}(2\widehat{P} - 1)^2 P(1 - P)$ $+ (2P - 1)^2 \widehat{P}(1 - \widehat{P})$ $+ \frac{4}{m} P \cdot \widehat{P}(1 - P)(1 - \widehat{P})]$
Compute variance for SR	$var(SR) = \frac{1}{m} SRI(1 - SRI)$
Compute DA statistic	$DA = \frac{SR - SRI}{\sqrt{var(SR) - var(SRI)}} \overset{a}{\sim} N(0, 1)$

The DA statistic is approximately distributed as standard normal, under the null hypothesis that the signs of the forecasts and the signs of the actual variables are independent.

4.2.7 Predictive Stochastic Complexity

In choosing the best neural network specification, one has to make decisions regarding lag length for each of the regressors, as well as the type of network to be used, the number of hidden layers, and the number of networks in each hidden layer. One can, of course, make a quick decision on the lag length by using the linear model as the benchmark. However, if the underlying true model is a nonlinear one being approximated by the neural network, then the linear model should not serve this function.

Kuan and Liu (1995) introduced the concept of predictive stochastic complexity (PSC), originally put forward by Rissanen (1986a, b), for selecting both the lag and neural network architecture or specification. The basic approach is to compute the average squared honest or out-of-sample prediction errors and choose the network that gives the smallest PSC within a class of models. If two models have the same PSC, the simpler one should be selected.

Kuan and Liu applied this approach to exchange rate forecasting. They specified families of different feedforward and recurrent networks, with differing lags and numbers of hidden units. They make use of random specification for the starting parameters for each of the networks and choose the one with the lowest out-of-sample error as the starting value. Then they use a Newton algorithm and compute the resulting PSC values. They conclude that nonlinearity in exchange rates may be exploited by neural networks to "improve both point and sign forecasts" [Kuan and Liu (1995), p. 361].

4.2.8 Cross-Validation and the .632 Bootstrapping Method

Unfortunately, many times economists have to work with time series lacking a sufficient number of observations for both a good in-sample estimation and an out-of-sample forecast test based on a reasonable number of observations.

The reason for doing out-of-sample tests, of course, is to see how well a model generalizes beyond the original training or estimation set or historical sample for a reasonable number of observations. As mentioned above, the recursive methodology allows only one out-of-sample error for each training set. The point of any out-of-sample test is to estimate the in-sample bias of the estimates, with a sufficiently ample set of data. By *in-sample bias* we mean the extent to which a model overfits the in-sample data and lacks ability to forecast well out-of-sample.

One simple approach is to divide the initial data set into k subsets of approximately equal size. We then estimate the model k times, each time leaving out one of the subsets. We can compute a series of mean squared error measures on the basis of forecasting with the omitted subset. For k equal to the size of the initial data set, this method is called *leave out one.* This method is discussed in Stone (1977), Djkstra (1988), and Shao (1995).

LeBaron (1998) proposes a more extensive bootstrap test called the 0.632 bootstrap, originally due to Efron (1979) and described in Efron and Tibshirani (1993). The basic idea, according to LeBaron, is to estimate the original in-sample bias by repeatedly drawing new samples from the original sample, with replacement, and using the new samples as estimation sets, with the remaining data from the original sample not appearing in the new estimation sets, as clean test or out-of-sample data sets. In each of the repeated draws, of course, we keep track of which data points are in the estimation set and which are in the out-of-sample data set. Depending on the draws in each repetition, the size of the out-of-sample data set will vary. In contrast to cross-validation, then, the 0.632 bootstrap test allows a randomized selection of the subsamples for testing the forecasting performance of the model.

The 0.632 bootstrap procedure appears in Table 4.7.[2]

---

[2]LeBaron (1998) notes that the weighting 0.632 comes from the probability that a given point is actually in a given bootstrap draw, $1 - [1 - (\frac{1}{n})]^n \approx 1 - e^{-1} = 0.632$.

TABLE 4.7. 0.632 Bootstrap Test for In-Sample Bias

Obtain mean squared error from full data set	$MSSE^0 = \frac{1}{n}\sum_{i=1}^{n}[y_i - \widehat{y}_i]^2$
Draw a sample of length n with replacement	z_1
Estimate coefficients of model	Ω^1
Obtain omitted data from full data set	$\widetilde{z}$
Forecast out-of-sample with coefficients Ω^1	$\widehat{\widetilde{z}}_1 = \widehat{\widetilde{z}}_1(\Omega^1)$
Calculate mean squared error for out-of-sample data	$MSSE^1 = \frac{1}{n_1}\sum_{i=1}^{n_1}\left[\widetilde{z}_1 - \widehat{\widetilde{z}}_1\right]^2$
Repeat experiment B times	
Calculate average mean squared error for B boostraps	$\overline{MSSE} = \frac{1}{B}\sum_{b=1}^{B} MSSE^b$
Calculate bias adjustment	$\varpi^{(0.632)} = 0.632\left[MSSE^0 - \overline{MSSE}\right]$
Calculate adjusted error estimate	$MSSE^{(0.632)} = 0.368 \cdot MSSE^0$ $+0.632 \cdot \overline{MSSE}$

In Table 4.7, $\overline{MSSE}$ is a measure of the average mean out-of-sample squared forecast errors. The point of doing this exercise, of course, is to compare the forecasting performance of two or more competing models, to compare $MSSE_i^{(0.632)}$ for models $i = 1, \ldots, m$. Unfortunately, there is no well-defined distribution of the $MSSE^{(0.632)}$, so we cannot test if $MSSE_i^{(0.632)}$ from model i is significantly different from $MSSE_j^{(0.632)}$ of model j. Like the Hannan-Quinn information criterion, we can use this for ranking different models or forecasting procedures.

4.2.9 Data Requirements: How Large for Predictive Accuracy?

Many researchers shy away from neural network approaches because they are under the impression that large amounts of data are required to obtain accurate predictions. Yes, it is true that there are more parameters to estimate in a neural network than in a linear model. The more complex the network, the more neurons there are. With more neurons, there are more parameters, and without a relatively large data set, degrees of freedom diminish rapidly in progressively more complex networks.

In general, statisticians and econometricians work under the assumption that the more observations the better, since we obtain more precise and accurate estimates and predictions. Thus, combining complex estimation methods such as the genetic algorithm with very large data sets makes neural network approaches very costly, if not extravagant, endeavors. By *costly*, we mean that we have to wait a long time to get results, relative to linear models, even if we work with very fast hardware and optimized or fast software codes. One econometrician recently confided to me that she stays with linear methods because "life is too short."

Yes, we do want a relatively large data set for sufficient degrees of freedom. However, in financial markets, working with time series, too much data can actually be a problem. If we go back too far, we risk using data that does not represent very well the current structure of the market. Data from the 1970s, for example, may not be very relevant for assessing foreign exchange or equity markets, since the market conditions of the last decade have changed drastically with the advent of online trading and information technology. Despite the fact that financial markets operate with long memory, financial market participants are quick to discount information from the irrelevant past. We thus face the issue of data quality when quantity is abundant.

Walczak (2001) has examined the issue of length of the training set or in-sample data size for producing accurate forecasts in financial markets. He found that for most exchange-rate predictions (on a daily basis), a maximum of two years produces the "best neural network forecasting model performance" [Walczak (2001), p. 205]. Walczak calls the use of data closer in time to the data that are to be forecast the *times-series recency effect*. Use of more recent data can improve forecast accuracy by 5% or more while reducing the training and development time for neural network models [Walczak (2001), p. 205].

Walczak measures the accuracy of his forecasts not by the root mean squared error criterion but by percentage of correct out-of-sample direction of change forecasts, or directional accuracy, taken up by Pesaran and Timmerman (1992). As in most studies, he found that single-hidden-layer neural networks consistently outperformed two-layer neural networks, and that they are capable of reaching the 60% accuracy threshold [Walczak (2001), p. 211].

Of course, in macro time series, when we are forecasting inflation or productivity growth, we do not have daily data available. With monthly data, ample degrees of freedom, approaching in sample length the equivalent of two years of daily data, would require at least several decades. But the message of Walczak is a good warning that too much data may be too much of a good thing.

4.3   Interpretive Criteria and Significance of Results

In the final analysis, the most important criteria rest on the questions posed by the investigators. Do the results of a neural network lend themselves to interpretations that make sense in terms of economic theory and give us insights into policy or better information for decision making? The goal of computational and empirical work is insight as much as precision and accuracy. Of course, how we interpret a model depends on why we are estimating the model. If the only goal is to obtain better, more accurate forecasts, and nothing else, then there is no hermeneutics issue.

We can interpret a model in a number of ways. One way is simply to simulate a model with the given initial conditions, add in some small changes to one of the variables, and see how differently the model behaves. This is akin to impulse-response analysis in linear models. In this approach, we set all the exogenous shocks at zero, set one of them at a value equal to one standard deviation for one period, and let the model run for a number of periods. If the model gives sensible and stable results, we can have greater confidence in the model's credibility.

We may also be interested in knowing if some or any of the variables used in the model are really important or statistically significant. For example, does unemployment help explain future inflation? We can simply estimate a network with unemployment and then prune the network, taking unemployment out, estimate the network again, and see if the overall explanatory power or predictive performance of the network deteriorates after eliminating unemployment. We thus test the significance of unemployment as an explanatory variable in the network with a likelihood ratio statistic. However, this method is often cumbersome, since the network may converge at different local optima before and after pruning. We often get the perverse result that a network actually improves after a key variable has been omitted.

Another way to interpret an estimated model is to examine a few of the partial derivatives or the effects of certain exogenous variables on the dependent variable. For example, is unemployment more important for explaining future inflation than the interest rate? Does government spending have a positive effect on inflation? With these partial derivatives, we can assess, qualitatively and quantitatively, the relative strength of how exogenous variables affect the dependent variable.

Again, it is important to proceed cautiously and critically. An estimated model, usually an overfitted neural network, for example, may produce partial derivatives showing that an increase in firm profits actually increases the risk of bankruptcy! In complex nonlinear estimation such an absurd possibility happens when the model is overfitted with too many parameters.

The estimation process should be redone, by pruning the model to a simpler network, to find out if such a result is simply a result of too few or too many parameters in the approximation, and thus due to misspecification.

Absurd results can also come from the lack of convergence, or convergence to a local optimum or saddle point, when quasi-Newton gradient-descent methods are used for estimation.

In assessing the common sense of a neural network model it is important to remember that the estimated coefficients or the weights of the network, which encompass the coefficients linking the inputs to the neurons and the coefficients linking the neurons to the output, do not represent partial derivatives of the output y with respect to each of the input variables. As was mentioned, the neural network estimation is nonparametric, in the sense that the coefficients do not have a ready interpretation as behavioral parameters. In the case of the pure linear model, of course, the coefficients and the partial derivatives are identical.

Thus, to find out if an estimated network makes sense, we can readily compute the derivatives relating changes in the output variable with respect to changes in several input variables. Fortunately, computing such derivatives is a relatively easy task. There are two approaches: analytical and finite-difference methods.

Once we obtain the derivatives of the network, we can evaluate their statistical significance by bootstrapping. We next take up the topics of analytical and finite differencing for obtaining derivatives, and bootstrapping for obtaining significance, in turn.

4.3.1 Analytic Derivatives

One may compute the analytic derivatives of the output y with respect to the input variables in a feedforward network in the following way. Given the network:

$$n_{k,t} = \omega_{k,0} + \sum_{i=1}^{i^*} \omega_{k,i} x_{i,t} \tag{4.26}$$

$$N_{k,t} = \frac{1}{1 + e^{-n_{i,t}}} \tag{4.27}$$

$$y_t = \gamma_0 + \sum_{k=1}^{k^*} \gamma_k N_{k,t} \tag{4.28}$$

the partial derivative of y_t with respect to $x_{i^*,t}$ is given by:

$$\frac{\partial y_t}{\partial x_{i^*,t}} = \sum_{k=1}^{k^*} \gamma_k N_{k,t} (1 - N_{k,t}) \omega_{k,i^*} \tag{4.29}$$

The above derivative comes from an application of the chain rule:

$$\frac{\partial y_t}{\partial x_{i^*,t}} = \sum_{k=1}^{k^*} \frac{\partial y_t}{\partial N_{k,t}} \frac{\partial N_{k,t}}{\partial n_{k,t}} \frac{\partial n_{k,t}}{\partial x_{i^*,t}} \tag{4.30}$$

and from the fact that the derivative of a logsigmoid function N has the following property:

$$\frac{\partial N_{k,t}}{\partial n_{k,t}} = N_{k,t}[1 - N_{k,t}] \tag{4.31}$$

Note that the partial derivatives in the neural network estimation are indexed by t. Each partial derivative is state-dependent, since its value at any time or observation index t depends on the index t values of the input variables, x_t. The pure linear model implies partial derivatives that are independent of the values of x. Unfortunately, with nonlinear models one cannot make general statements about how the inputs affect the output without knowledge about the values of x_t.

4.3.2   Finite Differences

A more common way to compute derivatives are finite-difference methods. Given a neural network function, $y = f(x), x = [x_1, \ldots, x_i, \ldots, x_{i^*}]$, one way to approximate $f_{i,t}$ is through the one-sided finite-difference formula:

$$\frac{\partial y}{\partial x_i} = \frac{f(x_1, \ldots, x_i + h_i, \ldots, x_{i^*}) - f(x_1, \ldots, x_i, \ldots, x_{i^*})}{h_i} \tag{4.32}$$

where the denominator h_i is set at $\max(\epsilon, \epsilon.x_i)$, with $\epsilon = 10^{-6}$.

Second-order partial derivatives are computed in a similar manner. Cross-partials are given by the formula:

$$\frac{\partial^2 y}{\partial x_i \partial x_j} = \frac{1}{h_j h_i} \left[\begin{array}{l} \{f(x_1,\ldots,x_i+h_i,x_j+h_j,\ldots,x_{i^*})-f(x_1,\ldots,x_i,\ldots,x_j+h_j,\ldots,x_{i^*})\} \\ -\{f(x_1,\ldots,x_i+h_i,x_j,\ldots,x_{i^*})-f(x_1,\ldots,x_i,\ldots,x_j,\ldots,x_{i^*})\} \end{array} \right] \tag{4.33}$$

while the direct second-order partials are given by:

$$\frac{\partial^2 y}{\partial x_i^2} = \frac{1}{h_i^2} \left(\begin{array}{l} f(x_1, \ldots, x_i + h_i, x_j, \ldots, x_{i^*}) - 2f(\ldots x_i, \ldots, x_j, \ldots, x_{i^*}) \\ +(x_1, \ldots, x_i - h_i, x_j, \ldots, x_{i^*}) \end{array} \right) \tag{4.34}$$

where $\{h_i, h_j\}$ are the step sizes for calculating the partial derivatives. Following Judd (1998), the step size $h_i = \max(\varepsilon x_i, \varepsilon)$, where the scalar ε is set equal to the value 10^{-6}.

4.3.3   Does It Matter?

In practice, it does not matter very much. Knowing the exact functional form of the analytical derivatives certainly provides accuracy. However, for more complex functional forms, differentiation becomes more difficult, and as Judd (1998, p. 38) points out, finite-difference methods avoid errors that may arise from this source.

Another reason to use finite-difference methods for computing the partial derivatives of a network is that one can change the functional form, or the number of hidden layers in the network, without having to derive a new expression. Judd (1998) points out that analytic derivatives are better considered only when needed for accuracy reasons, or as a final stage for speeding up an otherwise complete program.

4.3.4   MATLAB Example: Analytic and Finite Differences

To show how closely the exact analytical derivatives and the finite differences match numerically, consider the logsigmoid function of a variable x, $1/[1+\exp(-x)]$. Letting x take on values from -1 to $+1$ at grid points of .1, we can compute the analytical and finite differences for this interval with the following MATLAB program, which calls the program *myjacobian.m*:

```
x = -1:.1:1; % Define the range of the input variable
x = x';
y = inv(1+exp(-x)); % Calculate the output variable
yprime_exact = y .* (1-y); % Calculate the analytical derivative
fun = 'logsig'; % Define function
h = 10 * exp(-6); % Define h
rr = length(x);
for i = 1:rr, % Calculate the finite derivative
yprime_finite(i,:) = myjacobian(fun, x(i,:), h);
end
% Obtain the mean of the squared error
meanerrorsquared = mean((yprime_finite - yprime_exact).^ 2);
```

The results show that the mean sum of squared differences between the exact and finite difference solutions is indeed a very small value; to be exact, 5.8562e-007.

The function *myjacobian* is given by the following code:

```
function jac = myjacobian(fun, beta, lambda);
% computes the jacobian matrix from the function;
% inputs: function, beta, lambda
% output: jacobian
[rr k] = size(beta);
```

```
value0 = feval(fun,beta);
vec1 = zeros(1,k);
for i = 1:k,
  vec2 = vec1;
  vec2(i) = max(lambda, lambda *beta(i));
  betax = beta + vec2;
  value1 = feval(fun,betax);
  jac(i) = (value1 - value0) ./ lambda;
  end
```

4.3.5 Bootstrapping for Assessing Significance

Assessing the statistical significance of an input variable in the neural network processes is straightforward. Suppose we have a model with several input variables. We are interested, for example, in whether or not government spending growth affects inflation. In a linear model, we can examine the t statistic. With nonlinear neural network estimation, however, the number of network parameters is much larger. As was mentioned, likelihood ratio statistics are often unreliable.

A more reliable but time-consuming method is to use the boostrapping method originally due to Efron (1979, 1983) and Efron and Tibshirani (1993). This bootstrapping method is different from the .632 bootstrap method for in-sample bias. In this method, we work with the original date, with the full sample, $[y, x]$, obtain the best predicted value with a neural network, $\widehat{y}$, and obtain the set of residuals, $\widehat{e} = y - \widehat{y}$. We then randomly sample this vector, $\widehat{e}$, with replacement and obtain the first set of shocks for the first bootstrap experiment, $\widehat{e}^{b1}$. With this set of first randomly sampled shocks from the base of residuals, $\widehat{e}^{b1}$, we generate a new dependent variable for the first bootstrap experiment, $y^{b1} = \widehat{y} + \widehat{e}^{b1}$, and use the new data set $[y^{b1} \ x]$ to re-estimate a neural network and obtain the partial derivatives and other statistics of interest from the nonlinear estimation. We then repeat this procedure 500 or 1000 times, obtaining $\widehat{e}^{bi}$ and y^{bi} for each experiment, and redo the estimation. We then order the set of estimated partial derivatives (as well as other statistics) from lowest to highest values, and obtain a probability distribution of these derivatives. From this we can calculate bootstrap p-values for each of the derivatives, giving the probability of the null hypothesis that each of these derivatives is equal to zero.

The disadvantage of the bootstrap method, as should be readily apparent, is that it is more time-consuming than likelihood ratio statistics, since we have to resample from the original set of residuals and re-estimate the network 500 or 1000 times. However, it is generally more reliable. If we can reject the null hypothesis that a partial derivative is equal to zero, based on resampling the original residuals and re-estimating the model 500 or 1000 times, we can be reasonably sure that we have found a significant result.

4.4   Implementation Strategy

When we face the task of estimating a model, the preceding material indicates that we have a large number of choices to make at all stages of the process, depending on the weights we put on in-sample or out-of-sample performance and the questions we bring to the research. For example, do we take logarithms and first-difference the data? Do we deseasonalize the data? What type of data scaling function should we use: the linear function, compressing the data between zero or one, or another one? What type of neural network specification should we use, and how should we go about estimating the model? When we evaluate the results, which diagnostics should we take more seriously and which ones less seriously? Do we have to do out-of-sample forecasting with a split-sample or a real-time method? Should we use the bootstrap method? Finally, do we have to look at the partial derivatives?

Fortunately, most of these questions generally take care of themselves when we turn to particular problems. In general, the goal of neural network research is to evaluate its performance relative to the standard linear model, or in the case of classification, to logit or probit models. If logarithmic first-differencing is the norm for linear forecasting, for example, then neural networks should use the same data transformation. For deciding the lag structure of the variables in a time-series context, the linear model should be the norm. Usually, lag section is based on repeated linear estimation of the in-sample or training data set for different lag lengths of the variables, and the lag structure giving the lowest value of the Hannan-Quinn information criterion is the one to use.

The simplest type of scaling should be used first, namely, the linear [0,1] interval scaling function. After that, we can check the robustness of the overall results with respect to the scaling function. Generally, the simplest neural network alternative should be used, with a few neurons to start. A good start would be the simple feedforward model or the jump-connection network which uses a combination of the linear and logsigmoid connections.

For estimation, there is no simple solution; the genetic algorithm generally has to be used. It may make sense to use the quasi-Newton gradient-descent methods for a limited number of iterations and not wait for full converge, particularly if there are a large number of parameters.

For evaluating the in-sample criteria, the first goal is to see how well the linear model performs. We would like a linear model that looks good, or at least not too bad, on the basis of the in-sample criteria, particularly in terms of autocorrelation and tests of nonlinearity. Very poor performance on the basis of these tests indicates that the model is not well specified. So beating a poorly specified model with a neural network is not a big deal. We would like to see how well a neural network performs relative to the best specified linear model.

Generally a network model should do better in terms of overall explanatory power than a linear model. However, the acid test of performance is out-of-sample performance. For macro data, real-time forecasting is the sensible way to proceed, while split-sample tests are the obvious way to proceed for cross-section data.

For obtaining the out-of-sample forecasts with the network models, we recommend the thick model approach advocated by Granger and Jeon (2002). Since no one neural network gives the same results if the starting solution parameters or the scaling functions are different, it is best to obtain an ensemble of predictions each period and to use a trimmed mean of the multiple network forecasts for a thick model network forecast.

For comparing the linear and thick model network forecasts, the root mean squared error criteria and Diebold-Mariano tests are the most widely used for assessing predictive accuracy. While there is no harm in using the bootstrap method for assessing overall performance of the linear and neural net models, there is no guarantee of consistency between out-of-sample accuracy through Diebold-Mariano tests and bootstrap dominance for one method or the other. However, if the real world is indeed captured by the linear model, then we would expect that linear models would dominate the nonlinear network alternatives under the real-time forecasting and bootstrap criteria.

In succeeding chapters we will illustrate the implementation of network estimation for various types of data and relate the results to the theory of this chapter.

4.5 Conclusion

Evaluation of the network performance relative to the linear approaches should be with some combination of in-sample and out-of-sample criteria, as well as by common sense criteria. We should never be afraid to ask how much these models add to our insight and understanding. Of course, we may use a neural network simply to forecast or simply to evaluate particular properties of the data, such as the significance of one or more input variables for explaining the behavior of the output variable. In this case, we need not evaluate the network with the same weighting applied to all three criteria. But in general we would like to see a model that has good in-sample diagnostics also forecast out-of-sample well and make sense and add to our understanding of economic and financial markets.

4.5.1 MATLAB Program Notes

Many of the programs are available for web searches and are also embedded in popular software programs such as EViews, but several are not.

For in-sample diagnostics, for the Ljung-Box and McLeod-Li tests, the program *qstatlb.m* should be used. For symmetry, I have written *engleng.m*, and for normality, *jarque.m*. The Lee-White-Granger test is implemented with *wnntest1.m*, and the Brock-Deckert-Scheinkman test is given by *bds1.m*

For out-of-sample performance, the Diebold-Mariano test is given by *dieboldmar.m*, and the Pesaran-Timerman directional accuracy test is given by *datest.m*.

For evaluating first and second derivatives by finite differences, I have written *myjacobian.m* and *myhessian.m*.

4.5.2 Suggested Exercises

For comparing derivatives obtained by finite differences with exact analytical derivatives, I suggest again using the MATLAB Symbolic Toolbox. Write in a function that has an exact derivative and calculate the expression symbolically using *funtool.m*. Then create a function and find the finite-difference derivative with *myjacobian.m*.

Part II

Applications and Examples

5
Estimating and Forecasting with Artificial Data

5.1 Introduction

This chapter applies the models and methods presented in the previous chapters to artificially generated data. This is done to show the power of the neural network approach, relative to autoregressive linear models, for forecasting relatively complex, though artificial, statistical processes.

The primary motive for using artificial data is that there are no limits to the size of the sample! We can estimate the parameters from a training set with sufficiently large degrees of freedom, and then forecast with a relatively ample test set. Similarly, we can see how well the fit and forecasting performance of a given training and test set from an initial sample or realization of the true stochastic process matches another realization coming from the same underlying statistical generating process.

The first model we examine is the stochastic chaos (SC) model, the second is the stochastic volatility/jump diffusion (SVJD) model, the third is the Markov regime switching (MRS) model, the fourth is a volatility regime switching (VRS) model, the fifth is a distorted long-memory (DLM) model, and the last is the Black-Scholes options pricing (BSOP) model. The SC model is widely used for testing predictive accuracy of various forecasting models, the SVJD and VRS models are commonly used models for representing volatile financial time series, and the MRS model is used for analyzing GDP growth rates. The DLM model may be used to represent an economy subject to recurring bubbles. Finally, the BSOP model is the benchmark model for calculating the arbitrage-free prices for options, under the assumption of the log normal distribution of asset returns. This chapter shows how well neural networks, estimated with the hybrid global-search genetic algorithm and local gradient approach, approximate the data generated by these models relative to the linear benchmark model.

In some cases, the structure is almost linear, so that the network should not perform much better than the linear model — but it also should not perform too much worse. In one case, the model is simply a martingale, in which case the best predictor of y_{t+1} is y_t. Again, the linear and network models should not diverge too much in this case. We assume in each of these cases that the forecasting agent does not know the true structure. Instead, the agent attempts to learn the true data generating process from linear and nonlinear neural network estimation, and forecast on the basis of these two methods.

In each case, we work with stationary data. Thus, the variables are first-differenced if there is a unit root. While the Dickey-Fuller unit root tests, discussed in the previous chapter, are based on linear autoregressive processes, we use these tests since they are standard and routinely used in the literature.

When we work with neural networks and wish to compare them with linear autoregressive models, we normally want to choose the best network model relative to the best linear model. The best network model may well have a different lag structure than the best linear model. We should choose the best specifications for each model on the basis of in-sample criteria, such as the Hannan-Quinn information criterion, and then see which one does better in terms of out-of-sample forecasting performance, either in real-time or in bootstrap approaches, or both. In this chapter, however, we either work with univariate series generated with simple one-period lags or with a cross-section series. We simply compare the benchmark linear model against a simple network alternative, with the same lag structure and three neurons in one hidden layer, in the standard "plain vanilla" multilayer perceptron or feedforward network.

For choosing the best linear specification, we use an ample lag structure that removes traces of serial dependence and minimizes the Hannan-Quinn information criterion. To evaluate the linear model fairly against the network alternative, the lag length should be sufficient to remove any obvious traces of specification error such as serial dependence. Since the artificial data in this chapter are intended to replicate properties of higher-frequency daily data, we select a lag length of four, on the supposition that forecasters would initially use such a lag structure (representing a year for quarterly data, or almost a full business week for daily data) for estimation and forecasting.

5.2 Stochastic Chaos Model

The stochastic chaos (SC) model has the following representation:

$$y_t = 4 \cdot \varsigma_t \cdot y_{t-1} \cdot (1 - y_{t-1})$$

$$\varsigma_t \tilde{} U(0,1)$$

$$y_0 = .5 \tag{5.1}$$

The stochastic term ς_t is a draw from a random uniform distribution. The variable y_t depends on its own lag, y_{t-1}, as well as on (y_{t-1}), multiplied by a factor of 4. One realization appears in Figure 5.1. An easy code for generating this series is given by the following list of MATLAB commands:

```
T = 500;
z = rand(T,1);
y(1,:) = .5;
for i = 2:T, y(i,:) = 4 * z(i,:) * y(i-1,:) * (1-y(i-1,:)); end
```

Notice that there are periods of consistent high volatility followed by flat stable intervals, indicating a series of nonlinear events. We also see that the stochastic model generates only positive values, since the shock comes from a uniform distribution. Such a stochastic chaos process may be useful for modeling either implied volatility or observed volatility processes rather than rate-of-return processes in financial markets, since volatility processes have, by definition, positive values. Figure 5.1 pictures one such realization of a stochastic chaos process.

Chaos theory has been widely studied in applications to finance to see if there are hidden chaotic properties within financial market data. One of the properties of the stochastic process is that, for a given set of shocks $\{\varsigma_t\}$, the set of outcomes $\{y_t\}$ does not vary much, after a suitable interval, for any initial condition y_0, provided that $0 < y_0 < 1$. However, before that suitable interval has passed, the system dynamics vary quite a bit for the given set of shocks $\{\varsigma_t\}$. Figure 5.2 pictures three stochastic chaos processes for the same shocks, for $y_0 = [.001, .5, .99]$. We see that the dashed and dotted curves, processes generated by the initial conditions $y_0 = [.5, .99]$, converge after five periods, whereas the process generated by $y_0 = [.001]$ takes about 15 periods to converge to the same values as generated by $y_0 = [.5, .99]$. Thus, effects of the initial conditions wear off with different speeds and show different volatilities for the same sets of shocks and same laws of motion.

For values $y_0 = 0$ or $y_0 = 1$, of course, the process remains at zero, and for $y_0 < 0$ and $y_0 > 1$, the process diverges very quickly. Thus, the process

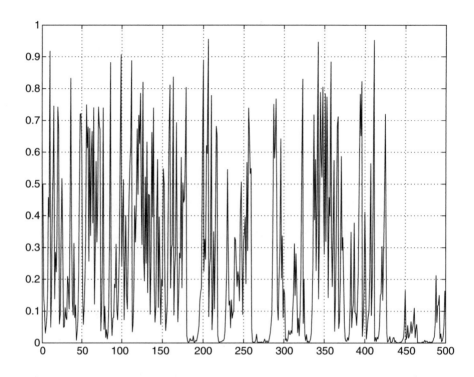

FIGURE 5.1. Stochastic chaos process

has an extreme sensitivity to very small changes in the initial condition, when the initial condition is in the neighborhood of zero or one.

5.2.1   In-Sample Performance

To fit the neural network and a linear model to this data set, we used both the genetic algorithm global search and the quasi-Newton local gradient methods. We withheld the last 20% (100 observations) as the out-of-sample test set, for real-time forecasting. We also used the bootstrap forecasting test.[1]

The in-sample performance of the linear model for the stochastic chaos model is summarized in Table 5.1.

Table 5.1 tells us that the linear model explains 29% of the variation of the in-sample data set, while the corresponding statistic of the

---

[1]The data generated by this model are estimated by neural network methods with the program *nnetjump.m*, available on the webpage of the author.

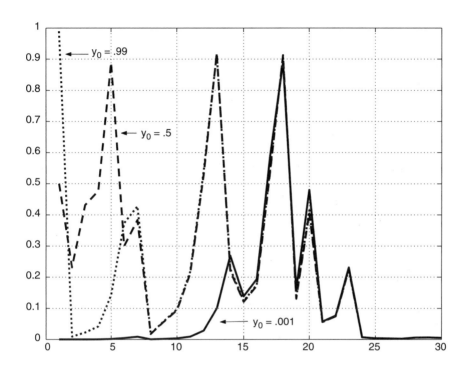

FIGURE 5.2. Stochastic chaos process for different initial conditions

TABLE 5.1. In-Sample Diagnostics: Stochastic
Chaos Model (Structure: 4 Lags, 3 Neurons)

Diagnostic	Linear Model (Network Model) Estimate
R^2	.29 (.53)
HQIF	1534 (1349)
L-B*	.251
M-L*	.0001
E-N*	.0000
J-B*	.55
L-W-G	1000
B-D-S*	.0000

* marginal significance levels

network model, appearing in parentheses, explains 53%. The Hannan-
Quinn information criterion favors, not surprisingly, the network model.
The significance test of the Q statistic shows that we cannot reject serial
independence of the regression residuals. By all other criteria, the linear

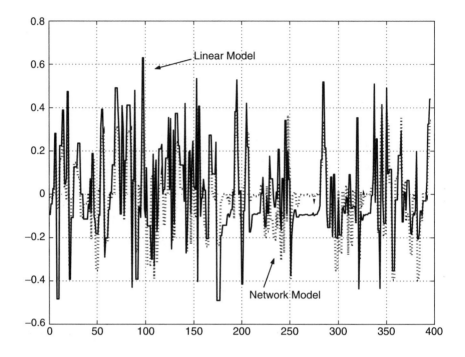

FIGURE 5.3. In-sample errors: stochastic chaos model

specification suffers from serious specification error. There is evidence of serial correlation in squared errors, as well as non-normality, asymmetry, and neglected nonlinearity in the residuals. Such indicators would suggest the use of nonlinear models as alternatives to the linear autoregressive structure.

Figure 5.3 pictures the error paths predicted by the linear and network models. The linear model errors are given by the solid curve and the network errors by dotted paths. As expected, we see that the dotted curves generally are closer to zero.

5.2.2   Out-of-Sample Performance

The path of the out-of-sample prediction errors appears in Figure 5.4. The solid path represents the forecast error of the linear model while the dotted curves are for the network forecast errors. This shows the improved performance of the network relative to the linear model, in the sense that its errors are usually closer to zero.

Table 5.2 summarizes the out-of-sample statistics. These are the root mean squared error statistics (RMSQ), the Diebold-Mariano statistics for lags zero through four (DM-0 to DM-4), the success ratio for percentage

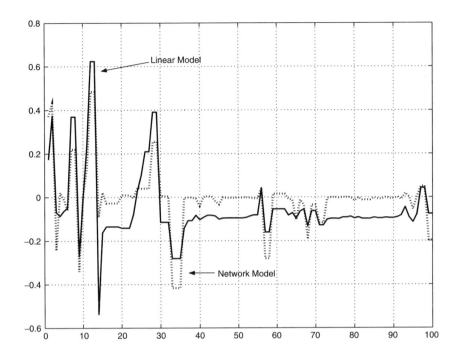

FIGURE 5.4. Out-of-sample prediction errors: stochastic chaos model

TABLE 5.2. Forecast Tests: Stochastic Chaos Model
(Structure: 5 Lags, 4 Neurons)

Diagnostic	Linear	Neural Net
RMSQ	.147	.117
DM-0[*]	—	.000
DM-1[*]	—	.004e-5
DM-2[*]	—	.032e-5
DM-3[*]	—	.115e-5
DM-4[*]	—	.209e-5
SR	1	1
B-Ratio	—	.872

[*] marginal significance levels

of correct sign predictions (SR), and the bootstrap ratio (B-Ratio), which
is the ratio of the network bootstrap error statistic to the linear bootstrap error measure. A value less than one, of course, represents a gain for
network estimation.

The results show that the root mean squared error statistic of the network model is almost 20% lower than that of the linear model. Not surprisingly, the Diebold-Mariano tests with lags zero through four are all significant. The success ratio for both models is perfect, since all of the returns in the stochastic chaos model are positive. The final statistic is the boot-strap ratio, the ratio of the network bootstrap error relative to the linear bootstrap error. We see that the network reduces the bootstrap error by almost 13%.

Clearly, if underlying data were generated by a stochastic process, networks are to be preferred over linear models.

5.3   Stochastic Volatility/Jump Diffusion Model

The SVJD model is widely used for representing highly volatile asset returns in emerging markets such as Russia or Brazil during periods of extreme macroeconomic instability. The model combines a stochastic volatility component, which is a time-varying variance of the error term, as well as a jump diffusion component, which is a Poisson jump process. Both the stochastic volatility component and the Poisson jump components directly affect the mean of the asset return process. They are realistic *para-metric* representations of the way many asset returns behave, particularly in volatile emerging-market economies.

Following Bates (1996) and Craine, Lochester, and Syrtveit (1999), we present this process in continuous time by the following equations:

$$\frac{dS}{S} = (\mu - \lambda \bar{k}) \cdot dt + \sqrt{V} \cdot dZ + k \cdot dq \qquad (5.2)$$

$$dV = (\alpha - \beta V) \cdot dt + \sigma_v \sqrt{V} \cdot dZ_v \qquad (5.3)$$

$$Corr(dZ, dZ_v) = \rho \qquad (5.4)$$

$$prob(dq = 1) = \lambda \cdot dt \qquad (5.5)$$

$$\ln(1 + k) \sim \phi(\ln[1 + \bar{k}] - .5\kappa, \kappa^2) \qquad (5.6)$$

where dS/S is the rate of return on an asset, μ is the expected rate of appreciation, λ the annual frequency of jumps, and k is the random per-centage jump conditional on the jump occurring. The variable $\ln(1 + k)$ is distributed normally with mean $\ln[1 + \bar{k}] - .5\kappa$ and variance κ^2. The symbol ϕ represents the normal distribution. The advantage of the continuous time representation is that the time interval can become arbitrarily smaller and approximate real time changes.

TABLE 5.3. Parameters for SVJD Process

Mean return	μ	.21
Mean volatility	α	.0003
Mean reversion of volatility	β	.7024
Time interval (daily)	dt	1/250
Expected jump	$\overline{k}$	.3
Standard deviation of percentage jump	κ	.0281
Annual frequency of jumps	λ	2
Correlation of Weiner processes	ρ	.6

The instantaneous conditional variance V follows a mean-reverting square root process. The parameter α is the mean of the conditional variance, while β is the mean-reversion coefficient. The coefficient σ_v is the variance of the volatility process, while the noise terms dZ and dZ_v are the standard continuous-time white noise Weiner processes, with correlation coefficient ρ.

Bates (1996) points out that this process has two major advantages. First, it allows systematic volatility risk, and second, it generates an "analytically tractable method" for pricing options without sacrificing accuracy or unnecessary restrictions. This model is especially useful for option pricing in emerging markets.

The parameters used to generate the SVJD process appear in Table 5.3.

In this model, S_{t+1} is equal to $S_t + [S_t \cdot (\mu - \lambda \overline{k})] \cdot dt$, and for a small value of dt will be unit-root nonstationary. After first-differencing, the model will be driven by the components of dV and $k \cdot dq$, which are random terms. We should not expect the linear or neural network model to do particularly well. Put another way, we should be suspicious if the network model significantly outperforms a rather poor linear model.

One realization of the SVJD process, after first-differencing, appears in Figure 5.5. As in the case of the stochastic chaos model, there are periods of high volatility followed by more tranquil periods. Unlike the stochastic chaos model, however, the periods of tranquility are not perfectly flat. We also notice that the returns in the SVJD model are both positive and negative.

5.3.1 In-Sample Performance

Table 5.4 gives the in-sample regression diagnostics of the linear model. Clearly, the linear approach suffers serious specification error in the error structure. Although the network multiple correlation coefficient is higher than that of the linear model, the Hannan-Quinn information criterion only slightly favors the network model. The slight improvement of the R^2 statistic does not outweigh by too much the increase in complexity due to

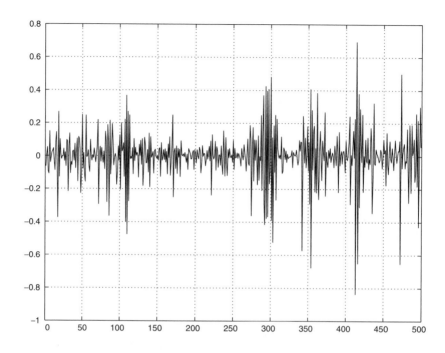

FIGURE 5.5. Stochastic volatility/jump diffusion process

TABLE 5.4. In-Sample Diagnostics: First-Differenced
SVJD Model (Structure: 4 Lags, 3 Neurons)

Diagnostic	Linear Model (Network Model) Estimate
R^2	.42 (.45)
HQIF	935 (920)
L-B*	.783
M-L*	.025
E-N*	.0008
J-B*	0
L-W-G	11
B-D-S*	.0000

* marginal significance levels

the larger number of parameters to be estimated. While the Lee-White-Granger test does not turn up evidence of neglected nonlinearity, the BDS test does. Figure 5.6 gives in-sample errors for the SVJD realizations. We do not see much difference.

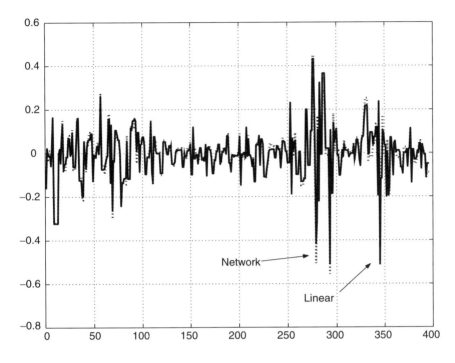

FIGURE 5.6. In-sample errors: SVJD model

5.3.2   Out-of-Sample Performance

Figure 5.7 pictures the out-of-sample errors of the two models. As expected, we do not see much difference in the two paths.

The out-of-sample statistics appearing in Table 5.5 indicate that the network model does slightly worse, but not significantly worse, than the linear model, based on the Diebold-Mariano statistic. Both models do equally well in terms of the success ratio for correct sign predictions, with slightly better performance by the network model. The bootstrap ratio favors the network model, reducing the error percentage of the linear model by slightly more than 3%.

5.4   The Markov Regime Switching Model

The Markov regime switching model is widely used in time-series analysis of aggregate macro data such as GDP growth rates. The basic idea of the

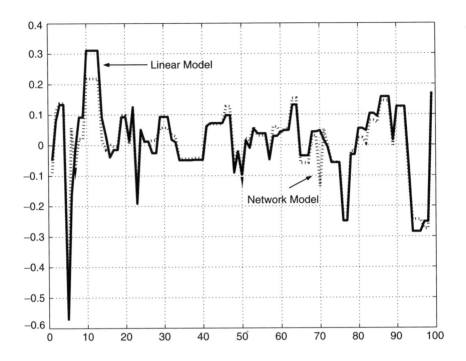

FIGURE 5.7. Out-of-sample prediction errors: SVJD model

TABLE 5.5. Forecast Tests: SVJD Model (Structure: 4 Lags, 3 Neurons)

Diagnostic	Linear	Neural Net
RMSQ	.157	.167
DM-0*	—	.81
DM-1*	—	.74
DM-2*	—	.73
DM-3*	—	.71
DM-4*	—	.71
SR	.646	.656
B-Ratio	—	.968

* marginal significance levels

regime switching model is that the underlying process is linear. However, the process follows different regimes when the economy is growing and when the economy is shrinking. Originally due to Hamilton (1990), it was applied to GDP growth rates in the United States.

Following Tsay (2002, p. 135–137), we simulate the following model representing the rate of growth of GDP for the U.S. economy for two states in the economy, S^1 and S^2:

$$x_t = c_c + \sum_{i-1}^{p} \phi_{1,i} x_{t-i} + \varepsilon_{1,i}, \ \varepsilon_1 \tilde{} \phi(0, \sigma_1^2), \text{if } S = S^1$$

$$= c_2 + \sum_{i-1}^{p} \phi_{2,i} x_{t-i} + \varepsilon_{2,i} \ \varepsilon_2 \tilde{} \phi(0, \sigma_2^2) \text{ if } S = S^2 \qquad (5.7)$$

where ϕ represents the Gaussian density function. These states have the following transition matrix, $\mathbf{P}$, describing the probability of moving from one state to the next, from time $(t-1)$ to time t:

$$\mathbf{P} = \begin{bmatrix} (S_{t,}^1 | S_{t-1,}^1) & (S_{t,}^1 | S_{t-1,}^2) \\ (S_{t,}^2 | S_{t-1,}^1) & (S_{t,}^2 | S_{t-1,}^2) \end{bmatrix} = \begin{bmatrix} (1-w_2) & w_2 \\ w_1 & (1-w_1) \end{bmatrix} \qquad (5.8)$$

The MRS model is essentially a combination of two linear models with different coefficients, with a jump or switch pushing the data-generating mechanism from one model to the other. So there is only a small degree of nonlinearity in this system. The parameters used for generating 500 realizations of the MRS model appear in Table 5.6.

Notice that in the specification of the transition probabilities, as Tsay (2002) points out, "it is more likely for the U.S. GDP to get out of a contraction period than to jump into one" [Tsay (2002), p. 137]. In our simulation of the model, the transition probability matrix is called from a uniform random number generator. If, for example, in state $S = S^1$, a random value of .1 is drawn, the regime will switch to the second state, $S = S^2$. If a value greater than .118 is drawn, then the regime will remain in the first state, $S = S^1$.

TABLE 5.6. Parameters for MRS Process

Parameter	State 1	State 2
c_i	.909	−.420
$\phi_{i,1}$	.265	.216
$\phi_{i,2}$	.029	.628
$\phi_{i,3}$	−.126	−.073
$\phi_{i,4}$	−.110	−.097
σ_i	.816	1.01
w_i	.118	.286

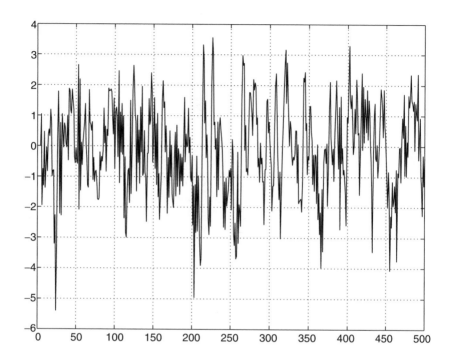

FIGURE 5.8. Markov switching process

The process $\{x_t\}$ exhibits periodic regime changes, with different dynamics in each regime or state. Since the representative forecasting agent does not know that the true data-generating mechanism for $\{x_t\}$ is a Markov regime switching model, a unit root test for this variable cannot reject an I(1) or nonstationary process. However, work by Lumsdaine and Papell (1997) and Cook (2001) has drawn attention to the bias of unit root tests when structural breaks take place. We thus approximate the process $\{x_t\}$ as a stationary process.

The underlying data-generating mechanism is, of course, near linear, so we should not expect great improvement from neural network approximation. One realization, for 500 observations, appears in Figure 5.8.

5.4.1 In-Sample Performance

Table 5.7 gives the in-sample regression diagnostics of the linear model. The linear regression model does not do a bad job, up to a point: there is no significant evidence of serial correlation in the residuals, and we cannot

TABLE 5.7. In-Sample Diagnostics: MRS
Model (Structure: 4 Lags, 3 Neurons)

Diagnostic	Linear Model (Network Model) Estimate
R^2	.35 (.38)
HQIF	3291 (3268)
L-B*	.91
M-L*	.0009
E-N*	.0176
J-B*	.36
L-W-G	13
B-D-S*	.0002

* marginal significance levels

reject normality in the distribution of the residuals. The BDS test shows
some evidence of neglected nonlinearity, but the LWG test does not.

Figure 5.9 pictures the error paths generated by the linear and neural net
models. While the overall explanatory power or R^2 statistic of the neural

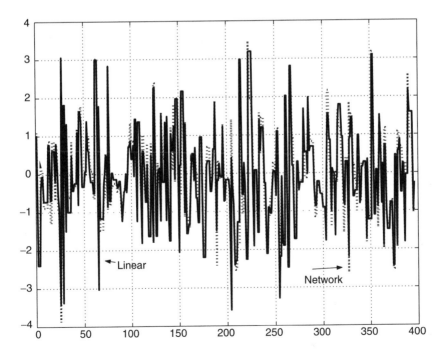

FIGURE 5.9. In-sample errors: MRS model

TABLE 5.8. Forecast Tests: MRS Model (Structure: 1 Lag, 3 Neurons)

Diagnostic	Linear	Neural Net
RMSQ	1.122	1.224
DM-0*	—	.27
DM-1*	—	.25
DM-2*	—	.15
DM-3*	—	.22
DM-4*	—	.24
SR	.77	.72
B-Ratio	—	.982

* marginal significance levels

net is slightly higher and the Hannan-Quinn information criterion indicates that the network model should be selected, there is not much noticeable difference in the two paths relative to the actual series.

5.4.2   Out-of-Sample Performance

The forecast statistics appear in Table 5.8. We see that the root mean squared error is slightly higher for the network, but the Diebold-Mariano statistics indicate that the difference in the prediction errors is not statistically significant. The bootstrap error ratio shows that the network model gives a marginal improvement relative to the linear benchmark.

The paths of the linear and network out-of-sample errors appear in Figure 5.10.

We see, not surprisingly, that both the linear and network models deliver about the same accuracy in out-of-sample forecasting. Since the MRS is basically a linear model with a small probability of a switch in the coefficients of the linear data-generating process, the network simply does about as well as the linear model.

What will be more interesting is the forecasting of the switches in volatility, rather than the return itself, in this series. We return to this subject in the following section.

5.5   Volatility Regime Switching Model

Building on the stochastic volatility and Markov regime switching models and following Tsay [(2002), p. 133], we use a simple autoregressive model with a regime switching mechanism for its volatility, rather than the return

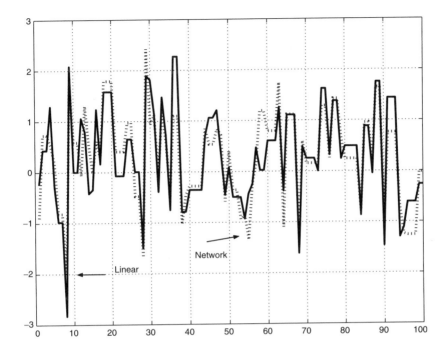

FIGURE 5.10. Out-of-sample prediction errors: MRS model

process itself. Specifically, we simulate the following model, similar to the one Tsay estimated as a process representing the daily log returns, including dividend payments, of IBM stock:[2]

$$r_t = .043 - .022r_{t-1} + \sigma_t + u_t \qquad (5.9)$$

$$u_t = \sigma_t \varepsilon_t, \ \varepsilon_t \tilde{\ } \phi(0,1) \qquad (5.10)$$

$$\sigma_t^2 = .098u_{t-1}^2 + .954\sigma_{t-1}^2 \ \text{if} \ u_{t-1} \leq 0$$

$$= .060 + .046u_{t-1}^2 + .8854\sigma_{t-1}^2 \ \text{if} \ u_{t-1} > 0 \qquad (5.11)$$

where $\phi(0,1)$ is the standard normal or Gaussian density. Notice that this VRS model will have drift in its volatility when the shocks are positive, but not when the shocks are negative. However, as Tsay points out, the

---

[2]Tsay (2002) omits the GARCH-in-Mean term $.5\sigma_t$ in his specification of the returns r_t.

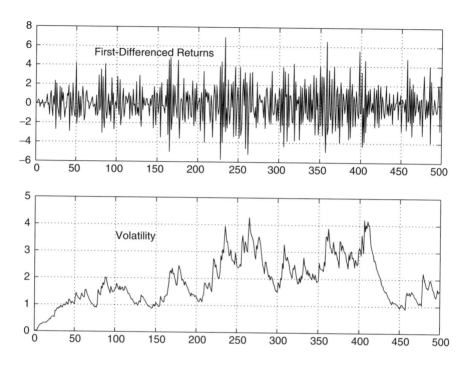

FIGURE 5.11. First-differenced returns and volatility of the VRS model

model essentially follows an IGARCH (integrated GARCH) when shocks are negative, since the coefficients sum to a value greater than unity.

Figure 5.11 pictures the first-differenced series of $\{r_t\}$, since we could not reject a unit-root process, as well as the volatility process $\{\sigma_t^2\}$.

5.5.1 In-Sample Performance

Table 5.9 gives the linear regression results for the returns. We see that the in-sample explanatory power of both models is about the same. While the tests for serial dependence in the residuals and squared residuals, as well as for symmetry and normality in the residuals, are not significant, the BDS test for neglected nonlinearity is significant. Figure 5.12 pictures the in-sample error paths of the two models.

5.5.2 Out-of-Sample Performance

Figure 5.13 and Table 5.10 show the out-of-sample performance of the two models. Again, there is not much to recommend the network model

TABLE 5.9. In-Sample Diagnostics: VRS
Model (Structure: 4 Lags, 3 Neurons)

Diagnostic	Linear Model (Network Model) Estimate
R^2	.422 (.438)
HQIF	3484 (3488)
L-B*	.85
M-L*	.13
E-N*	.45
J-B*	.22
L-W-G	6
B-D-S*	.07

* marginal significance levels

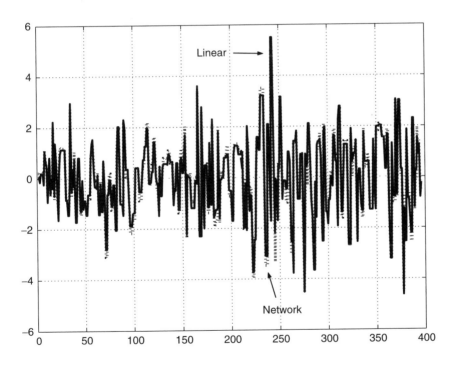

FIGURE 5.12. In-sample errors: VRS model

for return forecasting, but in its favor, it does not perform worse in any
noticeable way than the linear model.

While these results do not show overwhelming support for the superiority
of network forecasting for the volatility regime switching model, they do

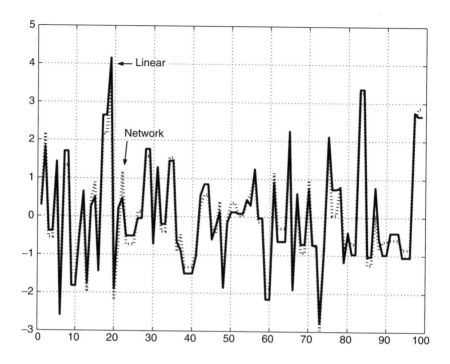

FIGURE 5.13. Out-of-sample prediction errors: VRS model

TABLE 5.10. Forecast Tests: VRS Model
(Structure: 4 Lags, 3 Neurons)

Diagnostic	Linear	Neural Net
RMSQ	1.37	1.38
DM-0*	—	.58
DM-1*	—	.58
DM-2*	—	.57
DM-3*	—	.56
DM-4*	—	.55
SR	.76	.76
B-Ratio	—	.99

* marginal significance levels

show improved out-of-sample performance both by the root mean squared
error and the bootstrap criteria. It should be noted once more that the
return process is highly linear by design. While the network does not do
significantly better by the Diebold-Mariano test, it does buy a forecasting
improvement at little cost.

5.6 Distorted Long-Memory Model

Originally put forward by Kantz and Schreiber (1997), the distorted long-memory (DLM) model was recently analyzed for stochastic neural network approximation by Lai and Wong (2001). The model has the following form:

$$y_t = x_{t-1}^2 x_t \tag{5.12}$$

$$x_t = .99x_{t-1} + \epsilon_t \tag{5.13}$$

$$\epsilon \sim N(0, \sigma^2) \tag{5.14}$$

Following Lai and Wong, we specify $\sigma = .5$ and $x_0 = .5$. One realization appears in Figure 5.14. It pictures a market or economy subject to bubbles. Since we can reject a unit root in this series, we analyze it in levels rather than in first differences.[3]

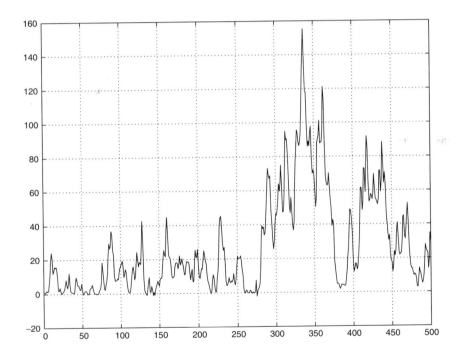

FIGURE 5.14. Returns of DLM model

---

[3]We note, however, the unit root tests are designed for variables emanating from a linear data-generating process.

TABLE 5.11. In-Sample Diagnostics: DLM Model (Structure: 4 Lags, 3 Neurons)

Diagnostic	Linear Model
R^2	.955 (.957)
HQIF	4900(4892)
L-B*	.77
M-L*	.0000
E-N*	.0000
J-B*	.0000
L-W-G	1
B-D-S*	.000001

* marginal significance levels

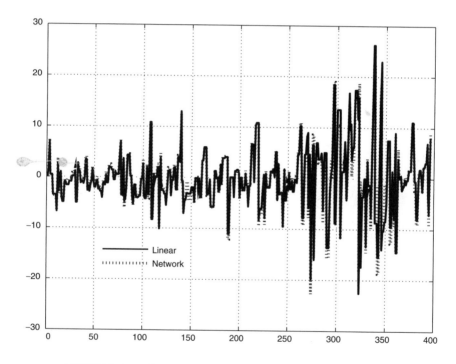

FIGURE 5.15. Actual and in-sample predictions: DLM model

5.6.1   In-Sample Performance

The in-sample statistics and time paths appear in Table 5.11 and Figure 5.15, respectively. We see that the in-sample power of the linear

TABLE 5.12. Forecast Tests: DLM Model
(Structure: 4 Lags, 3 Neurons)

Diagnostic	Linear	Neural Net
RMSQ	6.81	6.58
DM-0*	——	.09
DM-1*	——	.09
DM-2*	——	.05
DM-3*	——	.01
DM-4*	——	.02
SR	1	1
B-Ratio	——	.99

* marginal significance levels

model is quite high. The network model is slightly higher, and it is favored by the Hannan-Quinn criterion. Except for insignificant tests for serial independence, however, the diagnostics all indicate lack of serial independence, in terms of serial correlation of the squared errors, as well as non-normality, asymmetry, and neglected nonlinearity (given by the BDS test result). Since the in-sample predictions of the linear and neural network models so closely track the actual path of the dependent variable, we cannot differentiate the movements of these variables in Figure 5.15.

5.6.2 Out-of-Sample Performance

The relevant out-of-sample statistics appear in Table 5.12 and the prediction error paths are in Figure 5.16. We see that the root mean squared errors are significantly lower, while the success ratio for the sign predictions are perfect for both models. The network bootstrap error is also practically identical. Thus, the network gives a significantly improved performance over the linear alternative, on the basis of the Diebold-Mariano statistics, even when the linear alternative gives a very high in-sample fit.

5.7 Black-Sholes Option Pricing Model: Implied Volatility Forecasting

The Black-Sholes (1973) option pricing model is a well-known method for calculating arbitrage-free prices for options. As Peter Bernstein (1998) points out, this formula was widely in use by practitioners before it was recognized through publication in academic journals.

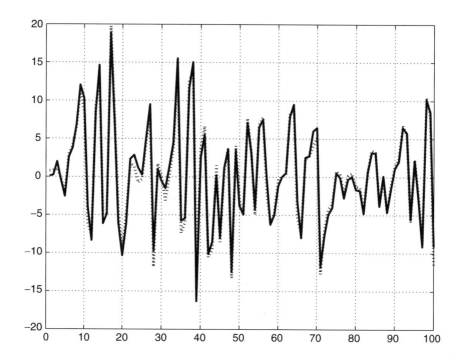

FIGURE 5.16. Out-of-sample prediction errors: DLM model

A *call option* is an agreement in which the buyer has the right, but not the obligation, to buy an asset at a particular strike price, X, at a preset future date. A *put option* is a similar agreement, with the right to sell an asset at a preset strike price. The options-pricing problem comes down to the calculation of an arbitrage-free price for the seller of the option. What price should the seller charge so that the seller will not systematically lose?

The calculation of the arbitrage-free price of the option in the Black-Sholes framework rests on the assumption of log-normal distribution of stock returns. Under this assumption, Black and Sholes obtained a closed-form solution for the calculation of the arbitrage-free price of an option. The solution depends on five variables: the market price of the underlying asset, S; the agreed-upon strike price, X; the risk-free interest rate, r_f; the maturity of the option, τ; and the annualized volatility or standard deviation of the underlying returns, σ. The maturity parameter τ is set at unity for annual, .25 for quarterly, .125 for monthly, and .004 for daily horizons.

The basic Black-Sholes formula yields the price of a European option. This type of option can be executed or exercised only at the time of maturity of the option. This formula has been extended to cover American options, in which the holder of the option may execute it at any time up to the expiration date of the option, as well as for options with ceilings or floors, which limit the maximum payout of the option.[4]

Options, of course, are widely traded on the market, so their price will vary from moment-to-moment. The Black-Sholes formula is particularly useful for calculating the issue price of new options. A newly issued option that is mispriced will be quickly arbitraged by market traders. In addition, the formula is often used for calculating the shadow price of different types of risk exposure. For example, a company expecting to receive revenue in British sterling over the next year, but that has costs in U.S. dollars, may wish to "price" their risk exposure. One price, of course, would be the cost of an option to cover their exposure to loss through a collapse of British sterling.[5]

Following Campbell, Lo, and MacKinlay (1997), the formula for pricing a call option is given by the following three equations:

$$C(S, X, \tau, \sigma) = S \cdot \Phi(d_1) - X \cdot \exp(-r \cdot \tau) \cdot \Phi(d_2) \qquad (5.15)$$

$$d_1 = \frac{\ln\left(\frac{S}{X}\right) + \left(r + \frac{\sigma^2}{2}\right)\tau}{\sigma\sqrt{\tau}} \qquad (5.16)$$

$$d_2 = \frac{\ln\left(\frac{S}{X}\right) + \left(r - \frac{\sigma^2}{2}\right)\tau}{\sigma\sqrt{\tau}} \qquad (5.17)$$

where $\Phi(d_1)$ and $\Phi(d_2)$ are the standard normal cumulative distribution functions of the variables d_1 and d_2. $C(S, X, \tau, \sigma)$ is the call option price of an underlying asset with a current market price S, with exercise price X, maturity τ, and annualized volatility σ.

Figure 5.17 pictures randomly generated values of S, X, r, τ, and σ as well as the calculated call option price from the Black-Scholes formula.

The call option data represent a random cross section for different types of assets, with different current market rates, exercise prices, risk-free rates, maturity horizons, and underlying volatility. We are not working with time-series observations in this approximation exercise. The goal of this exercise is to see how well a neural network, relative to a linear model, can approximate the underlying true Black-Sholes option pricing formula for predicting the not-call option price, given the observations on S, X, r, τ, and σ, but

---

[4]See Neftçi (2000) for a concise treatment of the theory and derivation of option-pricing models.

[5]The firm may also enter into a forward contract on foreign exchange markets. While preventing loss due to a collapse of sterling, the forward contract also prevents any gain due to an appreciation of sterling.

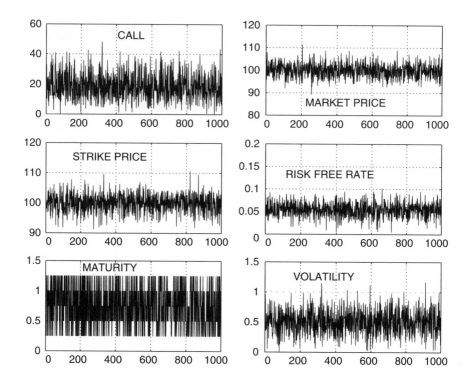

FIGURE 5.17.

rather the implied volatility from market data on option prices, as well as on S, X, r, τ.

Hutchinson, Lo, and Poggio (1994) have extensively explored how well neural network methods (including both radial basis and feedforward networks) approximate call option prices.[6] As these authors point out, were we working with time-series observations, it would be necessary to transform the independent variables S, X,and C into ratios, S_t/X_t and C_t/X_t.

5.7.1 In-Sample Performance

Table 5.13 gives the in-sample statistics. The R^2 statistic is relatively high, while all of the diagnostics are acceptable, except the Lee-White-Granger test for neglected nonlinearity.

---

[6]Hutchinson, Lo, and Poggio (1994) approximate the ratio of the call option price to the strike price, as a function of the ratio of the stock price to the strike price, and the time to maturity. They take the volatility and the risk-free rate of interest as given.

TABLE 5.13. In-Sample Diagnostics: BSOP
Model Structure:

Diagnostic	Linear Model (Network Model) Estimate
R^2	.91(.99)
HQIF	246($-$435)
L-B*	—
M-L*	—
E-N*	.22
J-B*	.33
L-W-G	997
B-D-S*	.47

* marginal significance levels

The in-sample error paths appear in Figure 5.18. The paths of both the
network and linear models closely track the actual volatility path. While
the R^2 for the network is slightly higher, there is not much appreciable
difference.

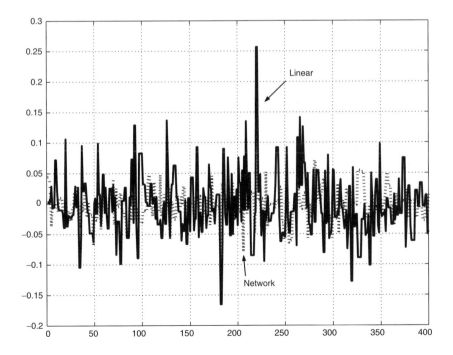

FIGURE 5.18. In-sample errors: BSOP model

TABLE 5.14. Forecast Tests: BSOP Model

Diagnostic	Linear	Neural Net
RMSQ	.0602	.0173
DM-0*	—	0
DM-1*	—	0
DM-2*	—	0
DM-3*	—	0
DM-4*	—	0
SR	1	1
B-Ratio	—	.28

* marginal significance levels

5.7.2   Out-of-Sample Performance

The superior out-of-sample performance of the network model over the linear model is clearly shown in Table 5.14 and in Figure 5.18. We see that the root mean squared error is reduced by more than 80% and the bootstrap error is reduced by more than 70%. In Figure 5.19, the network errors are closely distributed around zero, whereas there are large deviations with the linear approach.

5.8   Conclusion

This chapter evaluated the performance of alternative neural network models relative to the standard linear model for forecasting relatively complex artificially generated time series. We see that relatively simple feedforward neural nets outperform the linear models in some cases, or do not do worse than the linear models. In many cases we would be surprised if the neural networks did much better than the linear model, since the underlying data generating processes were almost linear.

The results of our investigation of these diverse stochastic experiments suggest that the real payoff from neural networks will come from volatility forecasting rather than pure return forecasting in financial markets, as we see in the high payoff from the implied volatility forecasting exercise with the Black-Sholes option pricing model. Since the neural networks never do appreciably worse than linear models, the only cost for using these methods is the higher computational time.

5.8.1   MATLAB Program Notes

The main script functions, as well as subprograms, are available on the website. The programs are *forecast_onevar_scmodel_new1.m* (for the stochastic

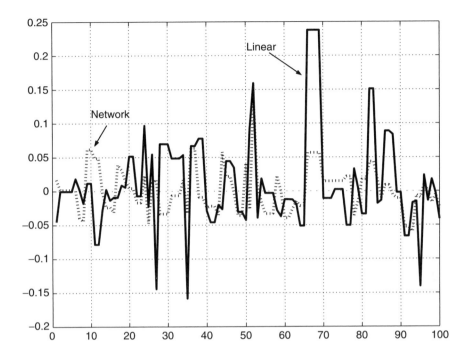

FIGURE 5.19. Out-of-sample prediction errors: BSOP model

chaos model), *forecast_onevar_svjdmodel_new1.m* (for the stochastic volatility jump diffusion model), *forecast_onevar_markovmodel_new1.m* (for the Markov regime switching model), and *forecast_onevar_dlm_new1.m* (for the distorted long-memory model).

5.8.2   *Suggested Exercises*

The programs in the previous section can be modified to generate alternative series of artificial data, extend the length of the sample, and modify the network models used for estimation and forecasting performance against the linear model. I invite the reader to continue these experiments with artificial data.

6

Times Series: Examples from Industry and Finance

This chapter moves the analysis away from artificially generated data to real-world data, to see how well the neural network model performs relative to the linear model. We focus on three examples: one from industry, the quantity of automobiles manufactured in the United States; one from finance, the spreads and the default rate on high-yield corporate bonds; and one from macroeconomics, forecasting inflation rates. In all three cases we use monthly observations.

Neural networks, of course, are routinely applied to forecasting very high-frequency data, such as daily exchange rates or even real-time share-market prices. However, in this chapter we show how the neural network performs when applied to more commonly used, and more widely accessible, data sets. All of the data sets are raw data sets, requiring adjustment for stationarity.

6.1 Forecasting Production in the Automotive Industry

The market for automobiles is a well-developed one, and there is a wealth of research on the theoretical foundations and the empirical behavior of this market. Since Chow (1960) demonstrated that this is one of the more stable consumer durable markets, empirical analysis has focused on improving the aggregate and disaggregated market forecasting with traditional time series as well as with pooled time-series cross-sectional methodologies, such as the study by McCarthy (1996).

The structure of the automobile market (for new vehicles) is recursive. Manufacturers evaluate and forecast the demand for the stock of automobiles, the number of retirements, and their market share. Adding a dose of strategic planning, they decide how much to produce. These decisions occur well before production and distribution take place. Manufacturers are providing a flow of capital goods to augment an existing stock. For their part, consumers decide at the time of purchase, based on their income, price, and utility requirements, what stock is optimal. To the extent that consumer decisions to expand the stock of the asset coincide with or exceed the amount of production by manufacturers, prices will adjust to revise the optimal stock and clear the market. To the extent they fall short, the number of retirements of automobiles will increase and the price of new vehicles will fall to clear the market. Chow (1960), Hess (1977), and McCarthy (1996) show how forecasting the demand in the markets is a sufficient proxy to modeling the optimal stock decision.

Both the general stability in the underlying market structure and the recursive nature of producer versus consumer decision making have made this market amenable to less complex estimation methods. Since research suggests this is precisely the kind of market in which linear time-series forecasting will perform rather well, it is a good place to test the usefulness of the alternative of neural networks for forecasting.[1]

6.1.1   The Data

We make use of quantity and price data for automobiles, as well as an interest rate and a disposable income as aggregate variables. The quantity variable represents the aggregate production of new vehicles, excluding heavy trucks and machinery, obtained from the Bureau of Economic Analysis of the Department of Commerce. The price variable is an index appearing in the Bureau of Labor Statistics. The interest rate argument is the home mortgage rate available from the Board of Governors of the U.S. Federal Reserve System, while the income argument is personal disposable income, also obtained from the Bureau of Economic Analysis of the Department of Commerce. Home mortgage rates were chosen as the relevant interest rate following Hess (1977), who shows that consumers consider housing and automobile decisions jointly. Personal disposable income was generated from consumption and savings data. The consumption series

---

[1]These points were made in a joint work with Gerald Nickelsburg. See McNelis and Nickelsburg (2002).

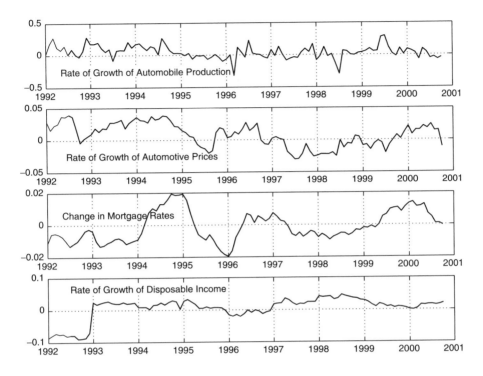

FIGURE 6.1. Automotive industry data

was the average over the quarter to reflect more accurately the permanent income concept.

Figure 6.1 pictures the evolution of the four variables we use in this example: annualized rates of change of the quantity and price indices obtained from the U.S. automotive industry, as well as the corresponding annual changes in the U.S. mortgage rates and the annualized rate of growth of U.S. disposable income.

We note some interesting features of the data: there has been no sharp rise in the rate of growth of prices since the mid-90s, while the peak year for automobile production growth took place between 1999 and 2000; and disposable income growth has been generally positive, with the exception of the recession at the end of the first Gulf War between 1992 and 1993.

Table 6.1 presents a statistical summary of these data.

We see that for the decade as a whole, there has been about a 4.5% annual growth in automobile production, whereas the price growth has been slightly less than 1% and disposable income growth has been about 0.5%. We also do not see a strong contemporaneous correlation between the variables. In fact, there are two "wrong" signs: a negative contemporaneous

TABLE 6.1. Summary of Automotive Industry Data

	Annualized Growth Rates: 1992–2001			
	Quantity	Price	Mortgage Rates	Disposable Income
Mean	0.0450	0.0077	−0.0012	0.0050
Std. Dev.	0.1032	0.0188	0.0092	0.0335
	Correlation Matrix			
	Quantity	Price	Mortgage Rates	Disposable Income
Quantity	1.0000			
Price	0.2847	1.0000		
Mortgage Rates	0.1248	0.1646	1.0000	0.2142
Disp. Income	−0.1703	−0.3304	0.2142	1.0000

correlation between disposable income growth and quantity growth, and a positive contemporaneous correlation between changes in mortgage rates and quantity growth.

6.1.2   Models of Quantity Adjustment

We use three models: a linear model, a smooth-transition regime switching model, and a neural network smooth-transition regime switching model (discussed in Section 2.5). We are working with monthly data. We are interested in the year-to-year changes in these data. When forecasting, we are interested in the annual or twelve-month forecast of the quantity of automobiles produced because investors are typically interested in the behavior of a sector over a longer horizon than one month or one quarter. Given the nature of lags in investment and time-to-build considerations, production over the next few months will have little to do with decisions made at time t.

Letting Q_t represent the quantity of automobiles produced at time t, we forecast the following variable:

$$\Delta_h q_{t+h} = q_{t+h} - q_t \tag{6.1}$$

$$q_t = \ln(Q_t) \tag{6.2}$$

where $h = 12$, for an annualized forecast with monthly data.

The dependent variable Δq_{t+h} depends on the following set of current variables $\mathbf{x}_t$

$$\mathbf{x}_t = [\Delta_{12} q_t, \Delta_{12} p_t, \Delta_{12} r_t, \Delta_{12} y_t] \tag{6.3}$$

$$\Delta_{12}p_t = \ln(P_t) - \ln(P_{t-12}) \tag{6.4}$$

$$\Delta_{12}r_t = \ln(R_t) - \ln(R_{t-12}) \tag{6.5}$$

$$\Delta_{12}y_t = \ln(Y_t) - \ln(Y_{t-12}) \tag{6.6}$$

where P_t, R_t, and Y_t signify the price index, the gross mortgage rate, and disposable income at time t. Although we can add further lags for Δq_t, we keep the set of regressions limited to the 12-month backward-looking horizon. The current value of Δq_t looks back over 12 months while the dependent variable looks forward over 12 months. We consider this a sufficiently ample lag structure. We also wish to avoid the problem of searching for different optimal lag structures for the three different models.

The linear model has the following specification:

$$\Delta q_{t+h} = \alpha \mathbf{x}_t + \eta_t \tag{6.7}$$

$$\eta_t = \epsilon_t + \gamma(L)\epsilon_{t-1} \tag{6.8}$$

$$\epsilon_t \sim N(0, \sigma^2) \tag{6.9}$$

The disturbance term η_t consists of a current period white-noise shock ϵ_t in addition to eleven lagged values of this shock, weighted by the vector γ. We explicitly model serial dependence as a moving average process since it is well known that whenever the forecast horizon exceeds the sampling interval, temporal dependence is induced in the disturbance term.

We compare this model with the smooth-transition regime switching (STRS) model and then with the neural network smooth-transition regime switching (NNSTRS) model. The STRS model has the following specification:

$$\Delta q_{t+h} = \Psi_t \alpha_1 \mathbf{x}_t + (1 - \Psi_t)\alpha_2 \mathbf{x}_t + \eta_t \tag{6.10}$$

$$\Psi_t = \Psi(\theta \cdot \Delta y_t - c) \tag{6.11}$$

$$= 1/[1 + \exp(\theta \cdot \Delta y_t - c)] \tag{6.12}$$

$$\eta_t = \epsilon_t + \gamma(L)\epsilon_{t-1} \tag{6.13}$$

$$\epsilon_t \sim N(0, \sigma^2) \tag{6.14}$$

where Ψ_t is a logistic or logsigmoid function of the rate of growth of disposable income, Δy_t, as well as the threshold parameter c and smoothness parameter θ. For simplicity, we set $c = 0$, thus specifying two regimes, one when disposable income is growing and the other when it is shrinking.

The NNSTRS model has the following form:

$$\Delta q_{t+h} = \alpha \mathbf{x}_t + \beta [\Psi_t G(\mathbf{x}_t; \alpha_1) + (1 - \Psi_t) H(\mathbf{x}_t; \alpha_2)] + \eta_t \qquad (6.15)$$

$$\Psi_t = \Psi(\theta \cdot \Delta y_t - c) \qquad (6.16)$$

$$= 1/[1 + \exp(\theta \cdot \Delta y_t - c)] \qquad (6.17)$$

$$G(\mathbf{x}_t; \alpha_1) = 1/[1 + \exp(-\alpha_1 \mathbf{x}_t)] \qquad (6.18)$$

$$H(\mathbf{x}_t; \alpha_2) = 1/[1 + \exp(-\alpha_2 \mathbf{x}_t)] \qquad (6.19)$$

$$\eta_t = \epsilon_t + \gamma(L)\epsilon_{t-1} \qquad (6.20)$$

$$\epsilon_t \sim N(0, \sigma^2) \qquad (6.21)$$

In the NNSTRS model, Ψ_t appears again as the transition function. The functions $G(\mathbf{x}_t; \alpha_1)$ and $H(\mathbf{x}_t; \alpha_2)$ are logsigmoid transformations of the exogenous variables $\mathbf{x}_t$, weighted by parameter vector α_1 in regime G and by vector α_2 in regime H. We note that the NNSTRS model has a direct linear component in which the exogenous variables are weighted by parameter vector α, and a nonlinear component, given by time-varying combinations of the two neurons, weighted by the parameter β.

The linear model is the simplest model, and the NNSTRS model is the most complex. We see that the NNSTRS nests the linear model. If the nonlinear regime switching effects are not significant, the parameter $\beta = 0$, so that it reduces to the linear model. The STRS model is almost linear, in the sense that the only nonlinear component is the logistic smooth-transition component Ψ_t. However, the STRS model nests the linear model only in a very special sense. With $\theta = c = 0$, $\Psi_t = .5$ for all t, so that the dependent variable is a linear combination of two linear models and thus a linear model. However, the NNSTRS does not nest the STRS model.

We estimate these three models by maximum likelihood methods. The linear model and the STRS models are rather straightforward to estimate. However, for the NNSTRS model the parameter set is larger. For this reason we make use of the hybrid evolutionary search (genetic algorithm) method and quasi-Newton gradient-descent methods. We then evaluate the relative performance of the three models by in-sample diagnostic checks, out-of-sample forecast accuracy, and the broader meaning and significance of the results.

6.1.3   In-Sample Performance

We first estimate the model for the whole sample period and assess the performance of the three models. Figure 6.2 pictures the errors of the models. The smooth lines represent the linear model, the dashed are for the STRS

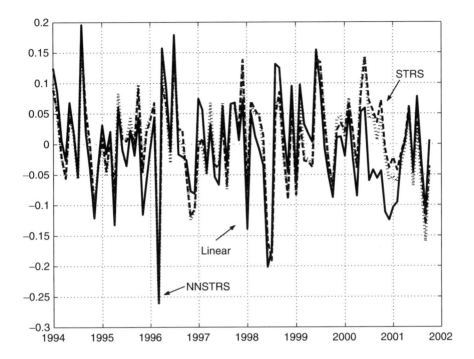

FIGURE 6.2. In-sample performance: rate of growth of automobile production

model, and the dotted curves are for the NNSTRS model. We see that the errors of the linear model are the largest, but they all are highly correlated with each other.

Table 6.2 summarizes the overall in-sample performance of the three models. We see that the NNSTRS model does not dominate the other STRS on the basis of the Hannan-Quinn selection criterion. For all three models we cannot reject serial independence, both in the residuals and in the squared residuals. Furthermore, the diagnostics on neglected nonlinearity are weakest on the linear model, but not by much, relative to the nonlinear models. All three models reject normality in the regression residuals.

6.1.4 Out-of-Sample Performance

We divided the sample in half and re-estimated the model in a recursive fashion for the last 53 observations. The real-time forecast errors appear in Figure 6.3. Again, the solid curves are for the linear errors, the dashed curves for the STRS model and the dotted curves are for the NNSTRS model. We see, for the most part, the error paths are highly correlated.

TABLE 6.2. In-sample Diagnostics of Alternative Models (Sample: 1992–2002, Monthly Data)

Diagnostics	Models		
	Linear	STRS	NNRS
SSE	0.615	0.553	0.502
RSQ	0.528	0.612	0.645
HQIF	−25.342	−22.714	−32.989
LB*	0.922	0.958	0.917
ML*	0.532	0.553	0.715
JB*	0.088	0.008	0.000
EN*	0.099	0.256	0.431
BDS*	0.045	0.052	0.051
LWG	0	0	0

*: prob value
NOTE:
SSE: Sum of squared errors
RSQ: R-squared
HIQF: Hannan-Quinn information criterion
LB: Ljung-Box Q statistic on residuals
ML: McLeod-Li Q statistic on squared residuals
JB: Jarque-Bera statistic on normality of residuals
EN: Engle-Ng test of symmetry of residuals
BDS:Brock-Deckert-Scheinkman test of nonlinearity
LWG: Lee-White-Granger test of nonlinearity

Table 6.3 summarizes the out-of-sample forecasting statistics of the three models. The root mean squared error statistics show the STRS model is the best, while the success ratio for correct sign prediction shows that the NNSTRS model is the winner. However, the differences between the two alternatives to the linear model are not very significant.

Table 6.3 has three sets of Diebold-Mariano statistics which compare, pair-wise, the three models against one another. Not surprisingly, given the previous information, the STRS and the NNSTRS errors are significantly better than the linear model, but they are not significantly different from each other.

6.1.5   Interpretation of Results

What do the models tell us in terms of economic understanding of the determinants of automotive production? To better understand the message of the models, we calculated the partial derivatives based on three states: the beginning of the sample, the mid-point, and the final observation. We also used the bootstrapping method to determine the statistical significance of these estimates.

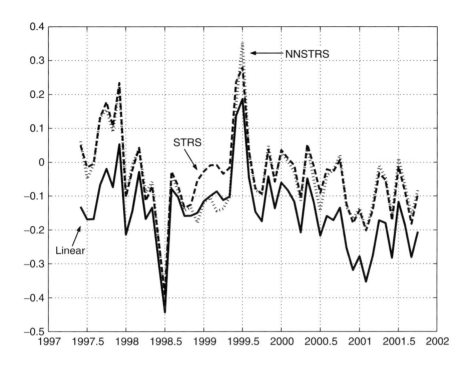

FIGURE 6.3.

TABLE 6.3. Out-of-Sample Forecasting Accuracy

Diagnostics	Models		
	Linear	STRS	NNSTRS
RMSQ	0.180	0.122	0.130
SR	0.491	0.679	0.698
Diebold-Mariano Test	Linear vs. STRS	Linear vs. NNSTRS	STRS vs. NNSTRS
DM-1*	0.000	0.000	0.941
DM-2*	0.000	0.002	0.899
DM-3*	0.000	0.005	0.874
DM-4*	0.000	0.009	0.857
DM-5*	0.000	0.013	0.853

*: prob value
RMSQ: Root mean squared error
SR: Success ratio on sign correct sign predictions
DM: Diebold-Mariano Test
(correction for autocorrelation, lags 1-5)

TABLE 6.4. Partial Derivatives of NNSTRS Model

Period	Arguments			
	Production	Price	Interest	Income
Mean	0.143	0.089	−0.450	0.249
1992	0.140	0.090	−0.458	0.249
1996	0.137	0.091	−0.455	0.248
2001	0.144	0.089	−0.481	0.250
Period	Statistical Significance of Estimates Arguments			
	Production	Price	Interest	Income
Mean	0.981	0.571	0.000	0.015
1992	0.968	0.558	0.000	0.001
1996	0.956	0.573	0.000	0.008
2001	0.958	0.581	0.000	0.008

The results appear in Table 6.4 for the NNSTRS model. We see that the partial derivatives of the mortgage rate and disposable income have the expected correct sign values and are statistically significant (based on bootstrapping) at the beginning, mid-point, and end-points of the sample, as well as for the mean values of the regressors. However, the partial derivatives of both the lagged production and the price are statistically significant. The message of the NNSTRS model is that aggregate macroeconomic variables are more important for predicting developments in automobile production than are price or lagged production developments within the industry itself.

The results from the STRS models are very similar, both in magnitude and tests of significance. These results appear in Table 6.5.

Finally, what information can we glean from the behavior of the smooth transition neurons in the two regime switching models? How do they behave relative to changes in disposable income? Figure 6.4 pictures the behavior of these three variables. We see that disposable income only becomes negative at the mid-point of the sample but at several points it is close to zero. The NNSTRS and STRS neurons give about equal weight to the growth/recession states, but the NNSTRS neuron shows slightly more volatility throughout the sample.

Given the superior performance of the STRS and NNSTRS models relative to the linear model, the information in Figure 6.4 indicates that most of the nonlinearity in the automotive industry has not experienced major switches in regimes. However, the neurons in both the STRS and NNSTRS model appear to detect nonlinearities which aid in forecasting performance.

TABLE 6.5. Partial Derivatives of STRS Model

Period	Arguments			
	Production	Price	Interest	Income
Mean	0.187	0.094	−0.448	0.296
1992	0.186	0.096	−0.449	0.291
1996	0.185	0.098	−0.450	0.286
2001	0.188	0.092	−0.448	0.299
Period	Statistical Significance of Estimates Arguments			
	Production	Price	Interest	Income
Mean	0.903	0.587	0.000	0.000
1992	0.905	0.575	0.000	0.000
1996	0.891	0.581	0.000	0.000
2001	0.893	0.589	0.000	0.000

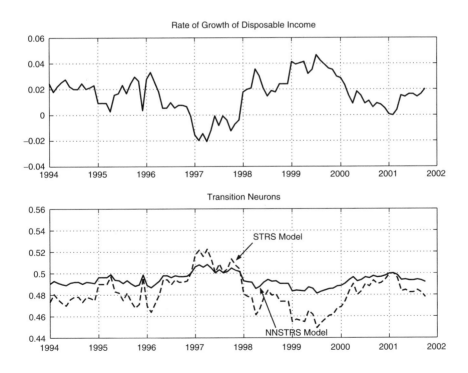

FIGURE 6.4. Regime transitions in STRS and NNSTRS models

6.2 Corporate Bonds: Which Factors Determine the Spreads?

The default rates of high-risk corporate bonds and the evolution of the spreads on the returns on these bonds, over ten-year government bond yields, appear in Figure 6.5.

What is most interesting about the evolution of both of these variables is the large upswing that took place at the time of the Gulf War recession in 1991, with the default rate appearing to lead the return spread. However, after 1992, both of these variables appear to move in tandem, without any clear lead or lag relation, with the spread variable showing slightly greater volatility after 1998. One fact emerges: the spreads declined rapidly in the early 90s, after the Gulf War recession, and started to increase in the late 1990s, after the onset of the Asian crisis in late 1997. The same is true of the default rates.

What is the cause of the decline in the spreads and the subsequent upswing of this variable? The process of financial market development may lead to increased willingness to take risk, as lenders attempt to achieve

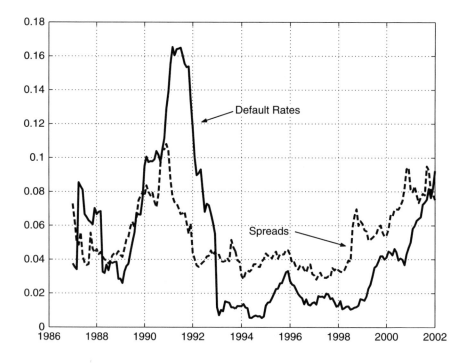

FIGURE 6.5. Corporate bond spreads and default rates

gains by broader portfolio diversification, which could explain a gradual decline, as lenders become less risk averse. Another factor may be the spillover effects from increases or decreases in the share market, as well as increased optimism or pessimism from the rate of growth of industrial production or from changes in confidence in the economy. These latter two variables represent business climate effects.

Collin-Dufresne, Goldstein, and Martin (2000) argue against macroeconomic determinants of credit spread changes in the U.S. corporate bond market. Their results suggest that the "corporate bond market is a segmented market driven by corporate bond specific supply/demand shocks" [Collin-Dufresne, Goldstein, and Martin (2000), p. 2]. In their view, the corporate default rates, representing "bond specific shocks," should be the major determinant of changes in spreads. They do find, however, that share market returns are negative and statistically significant determinants of the spreads. Like many previous studies, their analysis is based on linear regression methods.

6.2.1   The Data

We are interested in determining how these spreads respond to their own and each other's lagged values, to bond specific shocks such as default rates, as well as to key macroeconomic variables often taken as leading indicators of aggregate economic activity or the business climate: the real exchange rate, the index of industrial production (IIP), the National Association of Product Manufacturers' Index (NAPM), and the Morgan Stanley Capital International Index of the U.S. Share Market (MSCI). All of these variables, presented as annualized rates of change, appear in Figure 6.6.

Table 6.6 contains a statistical summary of these data. As in the previous example, we transform the spreads and default rates as annualized changes. We see in this table that over the 15-year period, 1987–2002, the average annualized change in the spread and the default rate is not very much. However, the volatility of the default rate is about three times higher. Of the macroeconomic and business climate indicators, we see that the largest growth, by far, took place in the MSCI index during this period of time. It also has the highest volatility.

The correlation matrix in Table 6.6 shows that the spreads are most highly negatively correlated with the NAPM index and most highly positively correlated with the default rate. In turn, the default rate is negatively correlated with changes in the index of industrial production (IIP).

6.2.2   A Model for the Adjustment of Spreads

We again use three models: a linear model, a smooth-transition regime switching model, and a neural network smooth-transition regime switching

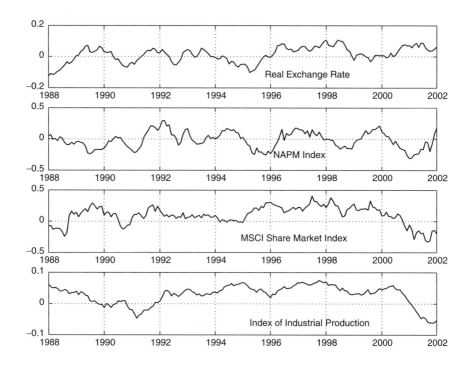

FIGURE 6.6. Annualized rates of change of macroeconomic indicators

TABLE 6.6. Annualized Changes of Financial Sector Indicators, 1987–2002

	Spread	Default Rate	Real. Ex. Rate	NAPM Index	MSCI Index	IIP
Mean	0.0021	0.0007	0.0129	−0.0181	0.1012	0.0288
Std. Dev.	0.0175	0.0363	0.0506	0.1334	0.1466	0.0317

Correlation Matrix

	Spread	Default Rates	Real. Ex. Rate	NAPM Index	MSCI Index	IIP
Spread	1					
Default Rate	0.3721	1				
Real. Ex. Rate	0.1221	0.0286	1			
NAPM Index	−0.6502	−0.2335	−0.0277	1		
MSCI Index	−0.0838	0.0067	0.2427	0.1334	1	
IIP	−0.1444	−0.4521	−0.1181	0.3287	0.4258	1

model (discussed in Section 2.5). Again we are working with monthly data, and we are interested in the year-on-year changes in these data. When forecasting the spread, financial market participants are usually interested in one-month or even shorter horizons.

Letting s_t represent the spread between corporate and U.S. government bonds at time t, we forecast the following variable:

$$\Delta s_{t+h} = s_{t+1} - s_t \tag{6.22}$$

where $h = 1$ for a one-period forecast with monthly data.

The dependent variable Δs_{t+h} depends on the following set of current variables $\mathbf{x}_t$

$$\mathbf{x}_t = [\Delta_{12}dr_t, \Delta s_t, \Delta_{12}rex_t, \Delta_{12}iip_t, \Delta_{12}msci_t, \Delta_{12}napm_t] \tag{6.23}$$

$$\Delta dr_t = dr_t - dr_{t-1} \tag{6.24}$$

$$\Delta_{12}rex_t = \ln(REX_t) - \ln(REX_{t-12}) \tag{6.25}$$

$$\Delta_{12}iip_t = \ln(IIP_t) - \ln(IIP_{t-12}) \tag{6.26}$$

$$\Delta_{12}msci_t = \ln(MSCI_t) - \ln(MSCI_{t-12}) \tag{6.27}$$

$$\Delta_{12}iip_t = \ln(NAPM_t) - \ln(NPAM_{t-12}) \tag{6.28}$$

where $\Delta_{12}dr_t$, Δs_t, $\Delta_{12}rex_t$, $\Delta_{12}iip_t$, $\Delta_{12}msci_t$, and $\Delta_{12}napm_t$ signify the currently observed changes in the default rate, the spreads, the index of industrial production, the MSCI stock index, and the NAPM index at time t. Since we work with monthly data, we use 12-month changes for the main macroeconomic indicators to smooth out seasonal factors.

The linear model has the following specification:

$$\Delta q_{t+h} = \alpha \mathbf{x}_t + \eta_t \tag{6.29}$$

$$\eta_t = \epsilon_t + \gamma(L)\epsilon_{t-1} \tag{6.30}$$

$$\epsilon_t \sim N(0, \sigma^2) \tag{6.31}$$

The disturbance term η_t consists of a current period white-noise shock ϵ_t in addition to eleven lagged values of this shock, weighted by the vector γ. We explicitly model serial dependence as a moving average process as in the previous case.

We compare this model with the smooth-transition regime switching (STRS) model and then with the neural network smooth-transition regime switching (NNSTRS) model. The STRS model has the following specification:

$$\Delta q_{t+h} = \Psi_t \alpha_1 \mathbf{x}_t + (1 - \Psi_t)\alpha_2 \mathbf{x}_t + \eta_t \tag{6.32}$$

$$\Psi_t = \Psi(\theta \cdot \Delta y_t - c) \tag{6.33}$$

$$= 1/[1 + \exp(\theta \cdot \Delta y_t - c)] \tag{6.34}$$

$$\eta_t = \epsilon_t + \gamma(L)\epsilon_{t-1} \tag{6.35}$$

$$\epsilon_t \sim N(0, \sigma^2) \tag{6.36}$$

where Ψ_t is a logistic or logsigmoid function of the rate of growth of disposable income, Δy_t, as well as the threshold parameter c and smoothness parameter θ. For simplicity, we set $c = 0$, thus specifying two regimes, one when disposable income is growing and the other when it is shrinking.

The NNSTRS model has the following form:

$$\Delta q_{t+h} = \alpha \mathbf{x}_t + \beta[\Psi_t G(\mathbf{x}_t; \alpha_1) + (1 - \Psi_t)H(\mathbf{x}_t; \alpha_2)] + \eta_t \tag{6.37}$$

$$\Psi_t = \Psi(\theta \cdot \Delta y_t - c) \tag{6.38}$$

$$= 1/[1 + \exp(\theta \cdot \Delta y_t - c)] \tag{6.39}$$

$$G(\mathbf{x}_t; \alpha_1) = 1/[1 + \exp(-\alpha_1 \mathbf{x}_t)] \tag{6.40}$$

$$H(\mathbf{x}_t; \alpha_2) = 1/[1 + \exp(-\alpha_2 \mathbf{x}_t)] \tag{6.41}$$

$$\eta_t = \epsilon_t + \gamma(L)\epsilon_{t-1} \tag{6.42}$$

$$\epsilon_t \sim N(0, \sigma^2) \tag{6.43}$$

6.2.3   In-Sample Performance

Figure 6.7 pictures the in-sample performance of the three models. We see that the linear predictions are clear outliers with respect to the two alternative models, especially at the time of the first Gulf War in late 1991.

The diagnostics appear in Table 6.7. We see a drastic improvement in performance as we abandon the linear model in favor of either the STRS or NNSTRS models. The Ljung-Box statistics indicate the presence of serial correlation in the linear model while we cannot reject independence in the alternatives. Both the Brock-Deckert-Scheinkman and Lee-White-Granger tests indicate the presence of neglected nonlinearities in the residuals of the linear model, but not in the residuals of the alternative models.

6.2.4   Out-of-Sample Performance

We again divided the sample in half and re-estimated the model in a recursive fashion for the last 86 observations. The real-time forecast errors appear in Figure 6.8. Again, the solid curves are for the linear errors, the dashed curves for the STRS model, and the dotted curves for the NNSTRS model. We see, for the most part, the error paths are highly correlated

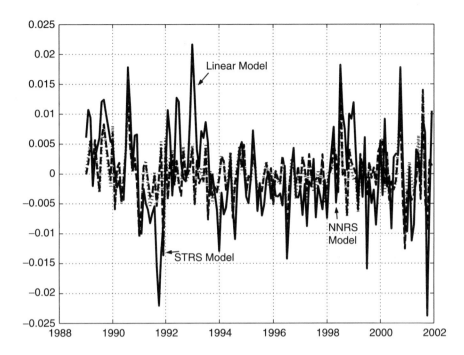

FIGURE 6.7. In-sample performance, change in bond spreads

with the two alternative models. However, large prediction error differences emerge in the mid-1990s, late 1990s, and late 2001.

Table 6.8 summarizes the out-of-sample forecasting statistics of the three models. The root mean squared error statistics show the STRS models as the best, while the success ratio for correct sign predictions (for the predicted change in the corporate bond spreads) shows that the STRS model is also the winner. However, the differences between the two alternatives to the linear model are not very significant.

Table 6.8 has three sets of Diebold-Mariano statistics which compare, pair-wise, the three models against one another. Again, the STRS and the NNSTRS errors are significantly better than the linear model, but they are not significantly different from each other.

6.2.5 Interpretation of Results

What do the models tell us in terms of economic understanding of the determinants of automotive production? To better understand the message of the models, we calculated the partial derivatives based on three states: the beginning of the sample, the mid-point, and the final observation. We also

TABLE 6.7. In-Sample Diagnostics of Alternative Models (Sample: 1988–2002, Monthly Data)

Diagnostics	Models		
	Linear	STRS	NNRS
SSE	0.009	0.003	0.003
RSQ	0.826	0.940	0.943
HQIF	−763.655	−932.234	−937.395
LB*	0.000	0.980	0.948
ML*	0.276	0.792	0.875
JB*	0.138	0.000	0.000
EN*	0.005	0.712	0.769
BDS*	0.000	0.338	0.297
LWG	798	0	0

*: prob value
NOTE:
SSE: Sum of squared errors
RSQ: R-squared
HIQF: Hannan-Quinn Information Criterion
LB: Ljung-Box Q statistic on residuals
ML: McLeod-Li Q statistic on squared residuals
JB: Jarque-Bera statistic on normality of residuals
EN: Engle-Ng test of symmetry of residuals
BDS:Brock-Deckert-Scheinkman test of nonlinearity
LWG: Lee-White-Granger test of nonlinearity

used the bootstrapping method to determine the statistical significance of these estimates.

The results appear in Table 6.9 for the NNSTRS model. We see significant and relatively strong persistence in the spread, in that current spreads have strong positive effects on the next period's spreads. We see that the effect of defaults is small and insignificant. The real exchange rate and industrial production effects are both positive and significant, while the effects of changes in the MSCI and NAPM indices are negative. In the NNSTRS model, however, the MSCI effect is not significant.

The message of the NNSTRS model is that aggregate macroeconomic variables are as important for predicting developments in spreads as are market-specific developments, since both the real exchange rate and changes in the NAPM, IIP, and lagged spreads play a significant role.

The results from the STRS models are very similar, both in magnitude and tests of significance. The only difference appears in the significance of the MSCI effect, which is significant in this model. This result is consistent with the findings of Collin-Dufresne, Goldstein, and Martin (2000). These results appear in Table 6.10.

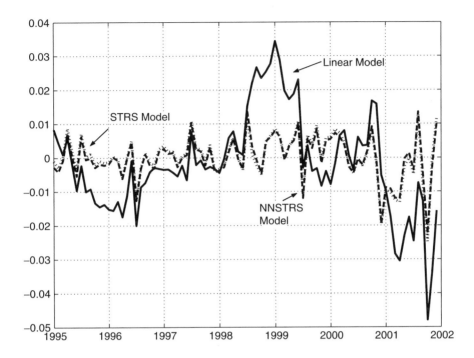

FIGURE 6.8. Forecasting performance of models

TABLE 6.8. Out-of-Sample Forecasting Accuracy

Diagnostics	Models		
	Linear	STRS	NNSTRS
RMSQ	0.015	0.006	0.007
SR	0.733	0.917	0.905
Diebold-Mariano Test	Linear vs. STRS	Linear vs. NNSTRS	STRS vs. NNSTRS
DM-1*	0.000	0.000	0.942
DM-2*	0.000	0.000	0.943
DM-3*	0.000	0.000	0.939
DM-4*	0.001	0.001	0.936
DM-5*	0.002	0.002	0.897

*: prob value
RMSQ: Root mean squared error
SR: Success ratio on correct sign predictions
DM: Diebold-Mariano Test
(correction for autocorrelation, lags 1–5)

TABLE 6.9. Partial Derivatives of NNSTRS Model

Period	Arguments					
	Default	Spread	REXR	IIP	MSCI	NAPM
Mean	0.033	0.771	0.063	0.134	−0.068	−0.066
1989	0.033	0.769	0.060	0.137	−0.065	−0.068
1996	0.030	0.777	0.071	0.128	−0.073	−0.061
2001	0.036	0.756	0.043	0.151	−0.053	−0.080

Statistical Significance of Estimates

Period	Arguments					
	Default	Spread	REXR	IIP	MSCI	NAPM
Mean	0.853	0.000	0.000	0.000	0.678	0.059
1989	0.844	0.000	0.000	0.000	0.688	0.055
1996	0.846	0.000	0.000	0.000	0.680	0.063
2001	0.848	0.000	0.000	0.000	0.684	0.055

TABLE 6.10. Partial Derivatives of STRS Model

Period	Arguments					
	Default	Spread	REXR	IIP	MSCI	NAPM
Mean	0.017	0.749	0.068	0.125	−0.139	−0.096
1989	0.010	0.752	0.070	0.128	−0.139	−0.098
1996	0.027	0.746	0.065	0.121	−0.138	−0.090
2001	−0.005	0.757	0.074	0.135	−0.140	−0.106

Statistical Significance of Estimates

Period	Arguments					
	Default	Spread	REXR	IIP	MSCI	NAPM
Mean	0.678	0.000	0.000	0.000	0.080	0.000
1989	0.699	0.000	0.000	0.000	0.040	0.000
1996	0.636	0.000	0.011	0.000	0.168	0.000
2001	0.693	0.000	0.000	0.000	0.057	0.000

Finally, we can ask what information we can glean from the behavior of the smooth transition neurons in the two regime switching models. How do they behave relative to changes in the IIP as the economy switches from growth to recession? Figure 6.9 pictures the behavior of these

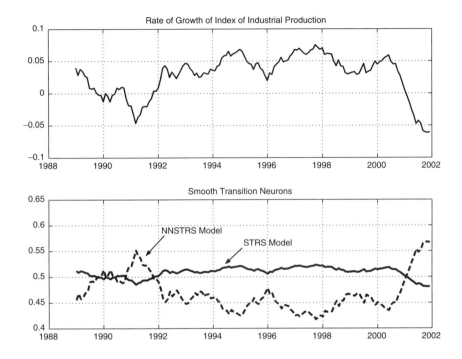

FIGURE 6.9. Regime transitions in STRS and NNSTRS models

three variables. We see sharper changes in the IIP index than in dispos-able income. The NNSTRS and STRS neurons give about equal weight to the growth/recession states, but the NNSTRS neuron shows slightly more volatility, and thus more information, throughout the sample about the likelihood of switching from one regime to another.

6.3   Conclusion

The examples we studied in this chapter are not meant, by any means, to be conclusive. The models are very simple and certainly capable of more elaborate extension, both in terms of the specification of the variables and in the specification of the nonlinear neural network alternatives to the linear model. However both of the examples illustrate the gains from using the nonlinear neural network specification, even in a simple alternative model. We get greater accuracy in forecasting and results with respectable in-sample diagnostics, which can lead to meaningful economic interpretation.

6.3.1   MATLAB Program Notes

The complete estimation program for the automobile industry and the spread forecasting exercises is called *carlos_may2004.m*. Subfunctions are *linearmodfun.m*, *nnstrsfun.m*, and *strsfun.m*, with the specification of a moving average process, for the linear, neural network smooth-transition regime switching, and smooth-transition regime switching models. The data for the corporate bond spreads are given in *carlos_spread_may2004_run1.mat*, while the automobile industry data are given in *jerryauto_may2004_run1.mat*.

6.3.2   Suggested Exercises

The reader is invited to modify the MATLAB programs and to forecast price adjustment, rather than quantity adjustment, in the automotive industry, and to forecast default rates, rather than corporate bond spreads, with the financial times-series data.

7

Inflation and Deflation: Hong Kong and Japan

This chapter applies neural network methods to the Hong Kong and Japanese experiences of inflation and deflation. Understanding the dynamics of inflation and how to forecast inflation more accurately is not simply of interest to policymakers at a central bank. Proper pricing of rates of return over the medium-term horizon requires accurate estimates of inflation in the coming quarters. Similarly, many decisions about lending or borrowing at short- or long-term interest rates requires a reasonable forecast of what succeeding short-term interest rates will be. These short-term interest rates, of course, will likely follow future inflationary developments, if the central bank is doing its job as a guardian of price stability. Forecasting inflation accurately means a better forecast of future interest rates and actions of the monetary authority.

Deflation poses a special problem. While at first glance the idea of falling prices appears to be good news, the zero lower bound on nominal interest rates means that real interest rates will start to rise sharply after the nominal interest rate hits its zero lower bound, if the deflation process continues. Rising real interest rates mean, of course, less investment and a fall in demand in the economy. Furthermore, a deflation process can generate self-fulfilling expectations. Once prices start to fall, people refrain from buying in the expectation that prices will continue to fall. The lack of buying, of course, causes prices to fall even more.

The dynamics of deflation raise many questions about the overall statistical process of inflation. When inflation is positive, we expect rising interest rates to reduce the inflationary pressures in the economy. However, in deflation, interest rates cannot fall below zero to reverse the deflationary pressure. There is an inherent asymmetry in the price adjustment process as we move from an inflationary regime to a deflationary regime. This is where we can expect nonlinear approximation methods to be of help.

While most studies of deflation have looked back to the Great Depression era, we have the more recent experiences of Hong Kong and Japan as new sources of information about how deflationary processes come about. While there has been great debate about the experiences of these countries, and no shortage of proposed policy remedies, there has been little examination of the inflationary/deflationary dynamics with nonlinear neural network approximation methods.

7.1 Hong Kong

Although much has been written (amid much controversy and debate) about deflation in Japan, which we discuss in Section 7.2, Hong Kong is of special interest. First, the usual response of expansionary monetary policy is not an option for Hong Kong, since its currency board arrangement precludes active policy directed at inflation or deflation. Second, Hong Kong is a smaller but much more open economy than Japan, and is thus more susceptible to external factors. Finally, Hong Kong, as a special administrative region, is in the process of increasing market integration with mainland China. However, there are some important similarities. Both Japan and Hong Kong have experienced significant asset-price deflation, especially in property prices, and more recently, negative output-gap measures.

Ha and Fan (2002) examined panel data for assessing price convergence between Hong Kong and mainland China. While convergence is far from complete, they showed that the pace has accelerated in recent years. However, comparing price dynamics between Hong Kong and Shenzhen, Schellekens (2003) argued that the role of price equalization as a source of deflation is minor, and contended that deflation is best explained by wealth effects.

Genberg and Pauwels (2003) found that both wages and import prices have significant causal roles, in addition to property rental prices. These three outperform measures of excess capacity as forcing variables for deflation. Razzak (2003) also called attention to the role of unit labor costs and productivity dynamics for understanding deflation. However, making use of a vector autoregressive model (VAR), Genberg (2003) also reported that external factors account for more than 50% of unexpected fluctuations in the real GDP deflator at horizons of one to two years.

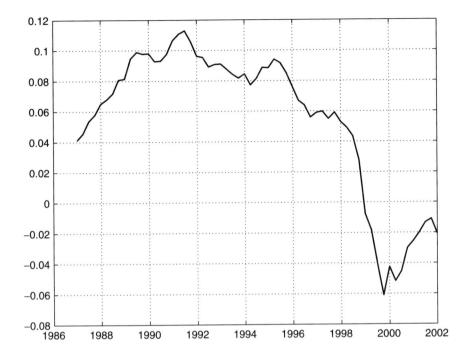

FIGURE 7.1. CPI inflation: Hong Kong

Most of these studies have relied on linear extensions and economet-ric implementation of the Phillips curve or New Keynesian Phillips curve. While such linear applications are commonly used and have been successful for many economies, we show in this chapter that a nonlinear smooth-transition neural network regime switching method outperforms the linear model on the basis of in-sample diagnostics and out-of-sample forecasting accuracy.

7.1.1 The Data

Figure 7.1 pictures the rate of inflation in Hong Kong. We see that the deflation process set in around 1998, reaching a rate of negative 6% by 1999. The country has not yet moved out of this pattern.

In this chapter, we examine the output gap, the rates of growth of import prices and unit labor costs, two financial sector indicators — the rates of growth of the Hang Seng index and residential property prices — and the price gap between Hong Kong and mainland China.

The output gap, which measures either excess demand or slack in the economy, comes from the World Economic Outlook of the IMF.

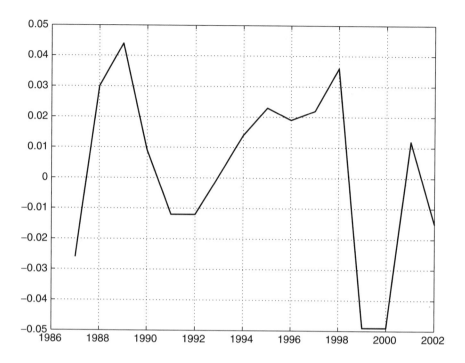

FIGURE 7.2. Output gap: Hong Kong

This variable was interpolated from annual to quarterly frequency. Figure 7.2 pictures the evolution of this variable. We see that measures of the output gap show that the economy has been well below potential for most of the time since late 1998.

The behavior of import prices and unit labor costs, both important for understanding the supply-side or costs factors of inflationary movements, shows considerably different patterns of volatility. Figure 7.3 pictures the rate of growth of import prices and Figure 7.4 shows the corresponding movement in labor costs. The collapse of import prices in the year 2001 is mainly due to the world economic downturn following the burst of the bubble in the high-technology sectors.

Figure 7.5 pictures the financial sector variables, the rates of growth of the share price index (the Hang Seng index), and the residential property price index. Not surprisingly, the growth rate of the share price index shows much more volatility than the corresponding growth rate of the property price index.

Finally, as a measure of structural market integration and price convergence with mainland China, we picture the evolution of a price gap.

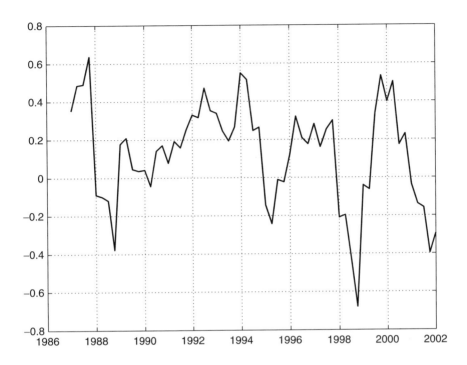

FIGURE 7.3. Rate of growth of import prices: Hong Kong

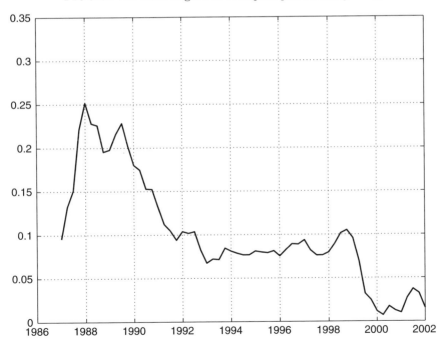

FIGURE 7.4. Rate of growth of unit labor costs: Hong Kong

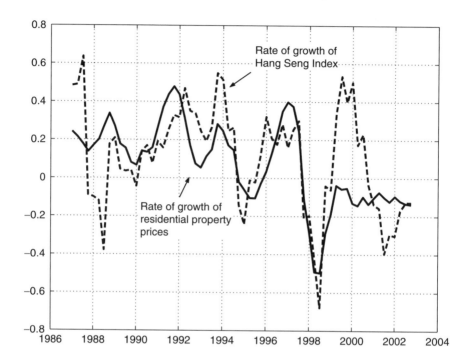

FIGURE 7.5. Financial market indicators: Hong Kong

The gap is defined as the logarithmic difference between the Hong Kong CPI and mainland China CPI. The latter is converted to the Hong Kong dollar basis using the market exchange rate. If there is significant convergence taking place, we expect a negative relationship between the price gap and inflation. If there is an unexpected and large price differential between Hong Kong and China, *ceteris paribus*, the inflation rate in Hong Kong should fall over time to close the gap. This variable appears in Figure 7.6.

Figure 7.6 shows that the price gap after 1998 is slowly but steadily falling. The jump in 1994 is due to the devaluation of the Chinese Renminbi against the U.S. dollar.

Table 7.1 contains a statistical summary of the data we use in our analysis. We use quarterly observations from 1985 until 2002. Table 7.1 lists the means, standard deviations, and contemporaneous correlations of annualized rates of inflation, the price and output gap measures, and the rates of growth of import prices, the property price index, the share price index, and unit labor costs.

The highest volatility rates (measured by the standard deviations of the annualized quarterly data) are for the rates of growth of the share market

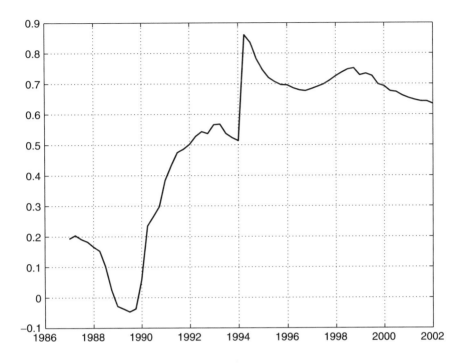

FIGURE 7.6. Price gap: Hong Kong and mainland China

and residential property price indices, as well as the price gap. However, the price gap volatility is due in large part to the once-over Renminbi devaluation in 1994.

Table 7.1 also shows that highest correlations of inflation are with rates of growth of unit labor costs and property prices, followed closely by the output gap. Finally, Table 7.1 shows a strong correlation between the growth rates of the share price and the residential property price indices.

In many studies relating to monetary policy and overall economic activity, bank lending appears as an important credit channel for assessing inflationary or deflationary impulses. Gerlach and Peng (2003) examined the interaction between banking credit and property prices in Hong Kong. They found that property prices are weakly exogenous and determine bank lending, while bank lending does not appear to influence property prices [Gerlach and Peng (2003), p. 11]. They argued that changes in bank lending cannot be regarded as the source of the boom and bust cycle in Hong Kong. They hypothesized that "changing beliefs about future economic prospects led to shifts in the demand for property and investments." With a higher inelastic supply schedule, this caused price swings, and with rising demand

TABLE 7.1. Statistical Summary of Data

Hong Kong Quarterly Data, 1985–2002

	Inflation	Price Gap	Output Gap	Imp Price Growth	Property Price Growth	HSI Growth	ULC Growth
Mean	0.055	0.511	0.004	0.023	0.088	0.127	0.102
Std. Dev.	0.049	0.258	0.024	0.051	0.215	0.272	0.062

Correlation Matrix

	Inflation	Price Gap	Output Gap	Imp Price Growth	Property Price Growth	HSI Growth	ULC Growth
Inflation	1.00						
Price Gap	−0.39	1.00					
Output Gap	0.56	−0.29	1.00				
Imp Price Growth	0.15	−0.37	0.05	1.00			
Property Price Growth	0.57	−0.42	0.36	0.23	1.00		
HSI Growth	0.06	−0.04	−0.15	0.43	0.56	1.00	
ULC Growth	0.59	−0.84	0.48	0.29	0.38	−0.09	1.00

for loans, "bank lending naturally responded" [Gerlach and Peng (2003), p. 11]. For this reason, we leave out the growth rate of bank lending as a possible determinant of inflation or deflation in Hong Kong.[1,2]

7.1.2 Model Specification

We draw upon the standard Phillips curve framework used by Stock and Watson (1999) for forecasting inflation in the United States. They define the inflation as an h-period ahead forecast. For our quarterly data set, we set $h = 4$ for an annual inflation forecast:

$$\pi_{t+h} = \ln(p_{t+h}) - \ln(p_t) \tag{7.1}$$

---

[1] In Japan, the story is different: banking credit and land prices show bidirectional causality or feedback. The collapse of land prices reduces bank lending, but the collapse of bank lending also leads to a fall in land prices. Hofmann (2003) also points out, with a sample of 20 industrialized countries, that "long run causality runs from property prices to bank lending" but short-run bidirectional causality cannot be ruled out.

[2] Goodhard and Hofmann (2003) support the finding of Gerlach and Peng with results from a wider sample of 12 countries.

We thus forecast inflation as an annual forecast (over the next four quarters), rather than as a one-quarter ahead forecast. We do so because policymakers are typically interested in the inflation prospects over a longer horizon than one quarter. For the most part, inflation over the next quarter is already in process, and changes in current variables will not have much effect at so short a horizon.

In this model, inflation depends on a set of current variables $\mathbf{x}_t$, including current inflation π_t, lags of inflation, and a disturbance term η_t. This term incorporates a moving average process with innovations ϵ_t, normally distributed with mean zero and variance σ^2 :

$$\pi_{t+h} = f(\mathbf{x}_t) + \eta_t \tag{7.2}$$

$$\pi_t = \ln(p_t) - \ln(p_{t-h}) \tag{7.3}$$

$$\eta_t = \epsilon_t + \gamma(\mathbf{L})\epsilon_{t-1} \tag{7.4}$$

$$\epsilon_t \sim N(0, \sigma^2) \tag{7.5}$$

where $\gamma(\mathbf{L})$ are lag operators. Besides current and lagged values of inflation, $\pi_t, \ldots, \pi_{t-k}$, the variables contained in $\mathbf{x}_t$ include measures of the output gap, y_t^{gap}, defined as the difference between actual output y_t and potential output y_t^{pot}, the (logarithmic) price gap with mainland China p_t^{gap}, the rate of growth of unit labor costs (ulc), and the rate of growth of import prices (imp). The vector $\mathbf{x}_t$ also includes two financial-sector variables: changes in the share price index (spi) and the residential property price index (rpi):

$$\mathbf{x}_t = [\pi_t, \ \pi_{t-1}, \pi_{t-2}, \ldots, \pi_{t-k}, y_t^{gap}, p_t^{gap}, \ldots,$$
$$\Delta_h ulc_t, \Delta_h imp_t, \Delta_h spi_t, \Delta_h rpi_t] \tag{7.6}$$

$$p_t^{gap} = p_t^{HK} - p_t^{CHINA} \tag{7.7}$$

The operator Δ_h for a variable z_t represents simply the difference over h periods. Hence $\Delta_h z_t = z_t - z_{t-h}$. The rates of growth of unit labor costs, the import price index, the share price index, and the residential property price index thus represent annualized rates of growth for $h = 4$ in our analysis. We do this for consistency with our inflation forecast, which is a forecast over four quarters. In addition, taking log differences over four quarters helps to reduce the influence of seasonal factors in the inflation process.

The disturbance term η_t consists of a current period shock ϵ_t in addition to lagged values of this shock. We explicitly model serial dependence, since it is well known that when the forecasting interval h exceeds the sampling interval (in this case we are forecasting for one year but we sample with quarterly observations), temporal dependence is induced in the disturbance term. For forecasting four quarters ahead with quarterly data, the error process is a third-order moving average process.

We specify four lags for the dependent variable. For quarterly data, this is equivalent to a 12-month lag for monthly data, used by Stock and Watson (1999) for forecasting inflation.

To make the model operational for estimation, we specify the following linear and neural network regime switching (NNRS) alternatives.

The linear model has the following specification:

$$\pi_{t+h} = \alpha \mathbf{x}_t + \eta_t \tag{7.8}$$

$$\eta_t = \epsilon_t + \gamma(\mathbf{L})\epsilon_{t-1} \tag{7.9}$$

$$\epsilon_t \sim N(0, \sigma^2) \tag{7.10}$$

We compare this model with the smooth-transition regime switching (STRS) model and then with the neural network smooth-transition regime switching (NNSTRS) model. The STRS model has the following specification:

$$\pi_{t+h} = \Psi_t \alpha_1 \mathbf{x}_t + (1 - \Psi_t)\alpha_2 \mathbf{x}_t + \eta_t \tag{7.11}$$

$$\Psi_t = \Psi(\theta \cdot \pi_{t-1} - c) \tag{7.12}$$

$$= 1/[1 + \exp(\theta \cdot \pi_{t-1} - c)] \tag{7.13}$$

$$\eta_t = \epsilon_t + \gamma(L)\epsilon_{t-1} \tag{7.14}$$

$$\epsilon_t \sim N(0, \sigma^2) \tag{7.15}$$

The transition function depends on the value of lagged inflation π_{t-1} as well as the parameter vector θ and threshold c, with $c = 0$. We use a logistic or logsigmoid specification for $\Psi(\pi_{t-1}; \theta, c)$.

We also compare the linear specification within a more general NNRS model:

$$\pi_{t+h} = \alpha \mathbf{x}_t + \beta\{[\Psi(\pi_{t-1}; \theta, c)]G(\mathbf{x}_t; \kappa)$$
$$+ [1 - \Psi(\pi_{t-1}; \theta, c)]H(\mathbf{x}_t; \lambda)\} + \eta_t \tag{7.16}$$

$$\eta_t = \epsilon_t + \gamma(\mathbf{L})\epsilon_{t-1} \tag{7.17}$$

$$\epsilon_t \sim N(0, \sigma^2) \tag{7.18}$$

The NNRS model is similar to the smooth-transition autoregressive model discussed in Franses and van Dijk (2000), originally developed by Teräsvirta (1994), and more generally discussed in van Dijk, Teräsvirta, and Franses (2000). The function $\Psi(\pi_{t-1}; \theta, c)$ is the transition function for two alternative nonlinear approximating functions $G(\mathbf{x}_t; \kappa)$ and $H(\mathbf{x}_t; \lambda)$.

The transition function is the same as the one used on the STRS model. Again, for simplicity we set the threshold parameter $c = 0$, so that the regimes divide into periods of inflation and deflation. As Franses and van Dyck (2000) point out, the parameter θ determines the smoothness of the change in the value of this function, and thus the transition from the inflation to deflation regime.

The functions $G(\mathbf{x}_t; \kappa)$ and $H(\mathbf{x}_t; \lambda)$ are also logsigmoid and have the following representations:

$$G(\mathbf{x}_t; \kappa) = \frac{1}{1 + \exp[-\kappa \mathbf{x}_t]} \tag{7.19}$$

$$H(\mathbf{x}_t; \lambda) = \frac{1}{1 + \exp[-\lambda \mathbf{x}_t]} \tag{7.20}$$

The inflation model in the NNRS model has a core linear component, including autoregressive terms, a moving average component, and a nonlinear component incorporating switching regime effects, which is weighted by the parameter β.

7.1.3   In-Sample Performance

Figure 7.7 pictures the in-sample paths of the regression errors. We see that there is little difference, as before, in the error paths of the two alternative models to the linear model.

Table 7.2 contains the in-sample regression diagnostics for the three models. We see that the Hannan-Quinn criteria only very slightly favors the STRS model over the NNRS model. We also see that the Ljung-Box, McLeod-Li, Brock-Deckert-Scheinkman, and Lee-White-Granger tests all call into question the specification of the linear model relative to the STRS and NNRS alternatives.

7.1.4   Out-of-Sample Performance

Figure 7.8 pictures the out-of-sample forecast errors of the three models. We see that the greatest prediction errors took place in 1997 (at the time of the change in the status of Hong Kong to a Special Administrative Region of the People's Republic of China).

The out-of-sample statistics appear in Table 7.3. We see that the root mean squared error statistic of the NNRS model is the lowest. Both the

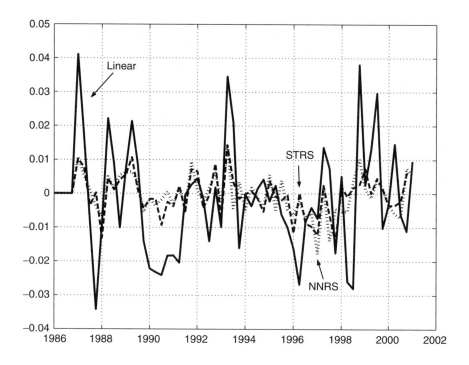

FIGURE 7.7. In-sample paths of estimation errors

STRS and NNRS models have much higher success ratios in terms of correct
sign predictions for the dependent variable, inflation. Finally, the Diebold-
Mariano statistics show that the NNRS prediction error path is significantly
different from that of the linear model and from the STRS model.

7.1.5   Interpretation of Results

The partial derivatives and their statistical significant values (based on
bootstrapping) appear in Table 7.4. We see that the statistically significant
determinates of inflation are lagged inflation, the output gap, the price
gap, changes in imported prices, the residential property price index, and
the Hang Seng index. Only unit labor costs are not significant. We also
see that the import price and price gap effects both have become more
important, with the import price derivative increasing from a value of .05
to a value of .13, from 1985 until 2002. This, of course, may reflect the
growing integration of Hong Kong both with China and with the rest of
the world. Residential property price effects have remained about the same.

TABLE 7.2. In-Sample Diagnostics of Alternative Models (Sample: 1985–2002, Quarterly Data)

Diagnostics	Models		
	Linear	STRS	NNRS
SSE	0.016	0.002	0.002
RSQ	0.965	0.983	0.963
HQIF	−230.683	−324.786	−327.604
LB*	0.105	0.540	0.316
ML*	0.010	0.204	0.282
JB*	0.282	0.856	0.526
EN*	0.441	0.792	0.755
BDS*	0.099	0.929	0.613
LWG	738	7	17

*: prob value
Note:
SSE: Sum of squared errors
RSQ: R-squared
HIQF: Hannan-Quinn information criterion
LB: Ljung-Box Q statistic on residuals
ML: McLeod-Li Q statistic on squared residuals
JB: Jarque-Bera statistic on normality of residuals
EN: Engle-Ng test of symmetry of residuals
BDS:Brock-Deckert-Scheinkman test of nonlinearity
LWG: Lee-White-Granger test of nonlinearity

For the sake of comparison, Table 7.5 pictures the corresponding information from the STRS model. The tests of significance are the same as in the NNRS model. The main differences are that the residential property price, import price, and output gap effects are stronger. But there is no discernible trend in the values of the significant partial derivatives as we move from the beginning of the sample period toward the end.

Figure 7.9 pictures the evolution of the smooth-transition neurons for the two models as well as the rate itself. We see that the neuron for the STRS model is more variable, showing a low probability of deflation in 1991, .4, but a much higher probability of deflation, .55, in 1999. The NNRS model has the probability remaining practically the same. This result indicates that the NNRS model is using the two neurons with equal weight to pick up nonlinearities in the overall inflation process independent of any regime change. If there is any slight good news for Hong Kong, the STRS model shows a very slight decline in the probability of deflation after 2000.

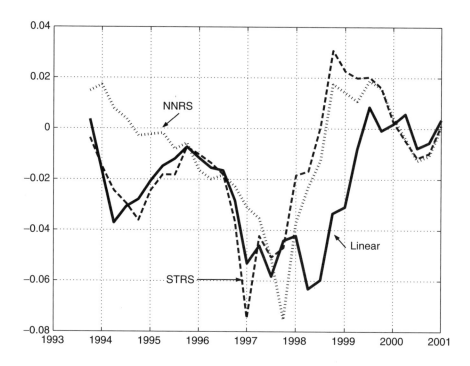

FIGURE 7.8. Out-of-sample prediction errors

TABLE 7.3. Out-of-Sample Forcasting Accuracy

Diagnostics	Models		
	Linear	STRS	NNRS
RMSQ	0.030	0.027	0.023
SR	0.767	0.900	0.867
Diebold-Mariano Test	Linear vs. STRS	Linear vs. NNRS	STRS vs. NNRS
DM-1*	0.295	0.065	0.142
DM-2*	0.312	0.063	0.161
DM-3*	0.309	0.031	0.127
DM-4*	0.296	0.009	0.051
DM-5*	0.242	0.000	0.002

*: prob value
RMSQ: Root mean squared error
SR: Success ratio on sign correct sign predictions
DM: Diebold-Mariano test
(correction for autocorrelation. lags 1–5)

TABLE 7.4. Partial Derivatives of NNSTRS Model

Period	Arguments						
	Inflation	Price Gap	Output Gap	Import Price	Res Prop Price	Hang Seng Index	Unit Labor Costs
Mean	0.300	−0.060	0.027	0.086	0.234	0.016	0.082
1985	0.294	−0.056	0.024	0.050	0.226	−0.015	0.072
1996	0.300	−0.060	0.027	0.091	0.235	0.020	0.084
2002	0.309	−0.067	0.032	0.130	0.244	0.053	0.093

Statistical Significance of Estimates

Period	Arguments						
	Inflation	Price Gap	Output Gap	Import Price	Res Prop Price	Hang Seng Index	Unit Labor Costs
Mean	0.000	0.000	0.015	0.059	0.000	0.032	0.811
1985	0.000	0.000	0.015	0.053	0.000	0.032	0.806
1996	0.000	0.000	0.013	0.034	0.000	0.029	0.819
2002	0.000	0.000	0.015	0.053	0.000	0.032	0.808

TABLE 7.5. Partial Derivatives of STRS Model

Period	Arguments						
	Inflation	Price Gap	Output Gap	Import Price	Res Prop Price	Hang Seng Index	Unit Labor Costs
Mean	0.312	−0.037	0.093	0.168	0.306	0.055	0.141
1985	0.295	−0.018	0.071	0.182	0.292	0.051	0.123
1996	0.320	−0.046	0.103	0.161	0.312	0.056	0.149
2002	0.289	−0.012	0.063	0.187	0.287	0.050	0.116

Statistical Significance of Estimates

Period	Arguments						
	Inflation	Price Gap	Output Gap	Import Price	Res Prop Price	Hang Seng Index	Unit Labor Costs
Mean	0.000	0.000	0.000	0.000	0.000	0.000	0.975
1985	0.000	0.000	0.000	0.000	0.000	0.000	0.964
1996	0.000	0.000	0.000	0.000	0.000	0.000	0.975
2002	0.000	0.000	0.000	0.000	0.000	0.000	0.966

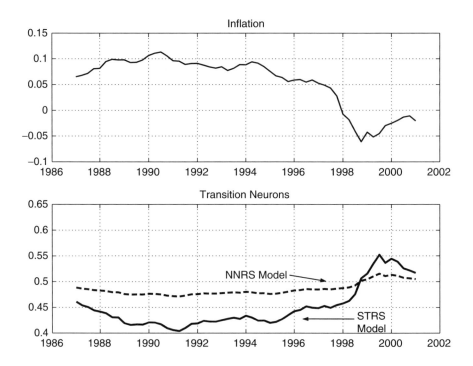

FIGURE 7.9. Regime transitions in STRS and NNRS models

7.2 Japan

Japan has been in a state of deflation for more than a decade. There is
no shortage of advice for Japanese policymakers from the international
community of scholars.

Krugman (1998) comments on this experience of Japan:

> Sixty years after Keynes, a great nation — a country with a stable and
> effective government, a massive net creditor, subject to none of the constraints
> that lesser economies face — is operating far below its productive capacity,
> simply because its consumers and investors do not spend enough. That should
> not happen; in allowing it to happen, and to continue year after year, Japan's
> economic officials have subtracted value from their nation and the world as a
> whole on a truly heroic scale [Krugman (1998), Introduction].

Krugman recommends expansionary monetary and fiscal policy to create inflation. However, Yoshino and Sakakibara have taken issue with
Krugman's remedies. They counter Krugman in the following way:

> Japan has reached the limits of conventional macroeconomic policies.
> Lowering interest rates will not stimulate the economy, because widespread

excess capacity has made private investment insensitive to interest rate changes. Increasing government expenditure in the usual way will have small effects because it will take the form of unproductive investment in the rural areas. Cutting taxes will not increase consumption because workers are concerned about job security and future pension and medical benefits [Yoshino and Sakakibara (2002), p. 110].

Besides telling us what will not work, Yoshino and Sakakibara offer alternative longer-term policy prescriptions, involving financial reform, competition policy, and the reallocation of public investment:

In order for sustained economic recovery to occur in Japan, the government must change the makeup and regional allocation of public investment, resolve the problem of nonperforming loans in the banking system, improve the corporate governance and operations of the banks, and strengthen the international competitiveness of domestically oriented companies in the agriculture, construction and service industries [Yoshino and Sakakibara (2002), p. 110].

Both Krugman and Yoshino and Sakakibara base their analyses and policy recommendations on analytically simple models, with reference to key stylized facts observed in macroeconomic data.

Svensson (2003) reviewed many of the proposed remedies for Japan, and put forward his own way. His "foolproof" remedy has three key ingredients: first, an upward-sloping price level target path set by the central bank; second, an initial depreciation followed by a "crawling peg;" and third, an exit strategy with abandonment of the peg in favor of inflation or price-level targeting when the price-level target path has been reached [Svensson (2003), p. 15]. Other remedies include a tax on money holding proposed by Goodfriend (2000) and Buiter and Panigirtzoglou (1999), as well as targeting the interest rate on long-term government bonds, proposed by Clouse *et al.* (2003) and Meltzer (2001).

The growth of low-priced imports from China has also been proposed as a possible cause of deflation in Japan (as in Hong Kong). McKibbin (2002) argued that monetary policy would be effective in Japan through yen depreciation. He argued for a combination of a fiscal contraction with a monetary expansion based on depreciation:

Combining a credible fiscal contraction that is phased in over three years with an inflation target would be likely to provide a powerful macroeconomic stimulus to the Japanese economy, through a weaker exchange rate and lower long term real interest rates, and would sustain higher growth in Japan for a decade [McKibbin (2002), p. 133].

In contrast to Krugman and Yoshino and Sakakibara, McKibbin based his analysis and policy recommendations on simulation of the calibrated G-cubed (Asia Pacific) dynamic general equilibrium model, outlined in McKibbin and Wilcoxen (1998).

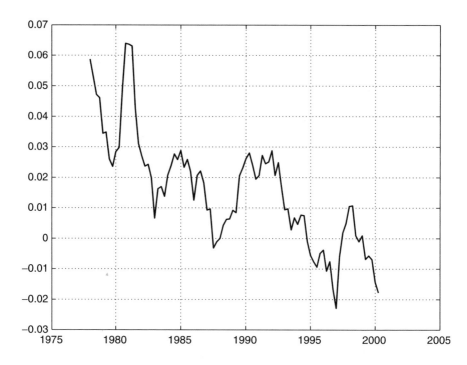

FIGURE 7.10. CPI inflation: Japan

Sorting out the relative importance of monetary policy, stimulus packages that affect overall demand (measured by the output gap), and the contributions of unit labor costs, falling imported goods prices, and financial-sector factors coming from the collapse of bank lending and asset-price deflation (measured by the negative growth rates of share price and land price indices) is no easy task. These variables display considerable volatility, and the response of inflation to these variables is likely to be asymmetric.

7.2.1 The Data

Figure 7.10 pictures the CPI inflation rate for Japan. We see that deflation set in after 1995, with a slight recovery from deflation in 1998.

Figure 7.11 pictures the output gap, while Figures 7.12 and 7.13 contain the rate of growth of the import price index and unit labor costs. We see that the collapse of excess demand, measured as a positive output gap, goes hand-in-hand with the onset of deflation. Unit labor costs also switched from positive to negative growth rate at the same time. However there is no noticeable collapse in the import price index at the time of the deflation.

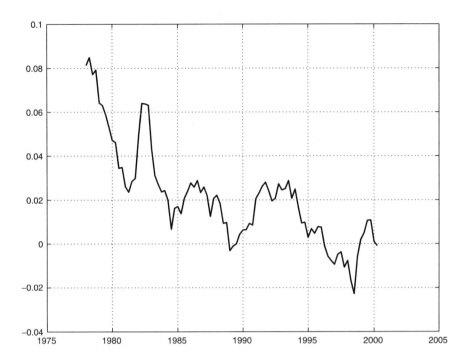

FIGURE 7.11. Output gap: Japan

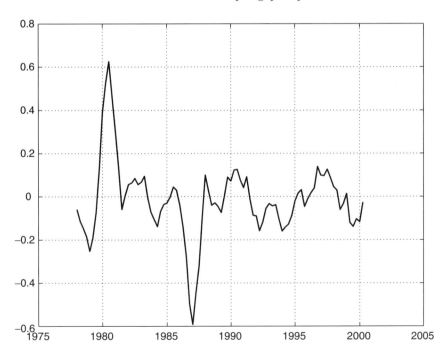

FIGURE 7.12. Rate of growth of import prices: Japan

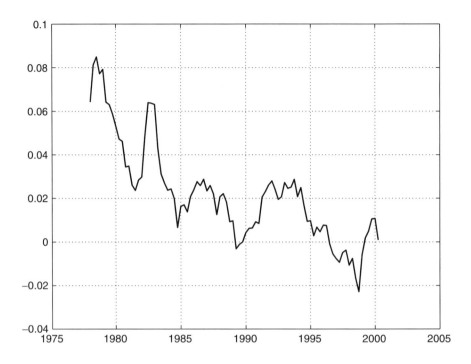

FIGURE 7.13. Rate of growth of unit labor costs: Japan

Figure 7.14 pictures the rate of growth of two financial market indicators: the Nikkei index and the land price index. We see that the volatility of the rate of growth of the Nikkei index is much greater than that of the land price index.

Figure 7.15 pictures the evolution of two indicators of monetary policy: the Gensaki interest rate and the rate of growth of bank lending. The Gensaki interest rate is considered the main interest for interpreting the stance of monetary policy in Japan. The rate of growth of bank lending is, of course, an indicator of how banks may thwart expansionary monetary policy by reducing their lending. We see the sharp collapse of the rate of growth of bank lending at about the same time the Bank of Japan raised the interest rates at the beginning of the 1990s. The well-documented action was an attempt by the Bank of Japan to burst the bubble in the stock market. Figure 7.14, of course, shows that the Bank of Japan did indeed succeed in bursting this bubble. After that, however, overall demand showed a steady decline.

Table 7.6 gives a statistical summary of the data we have examined. The highest volatility rates (measured by the standard deviations of the

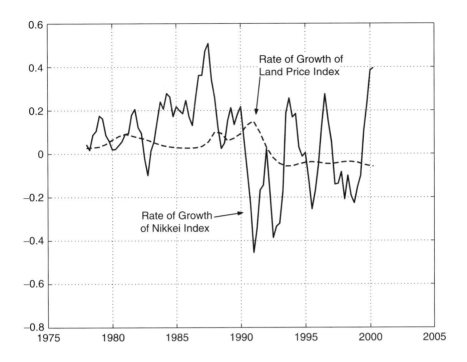

FIGURE 7.14. Financial market indicators: Japan

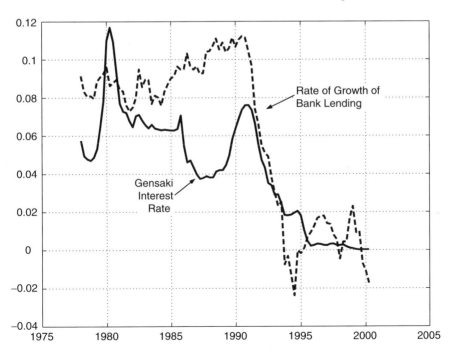

FIGURE 7.15. Monetary policy indicators: Japan

TABLE 7.6. Statistical Summary of Data

	Inflation	Gensaki	Y-gap	Imp Growth	Ulo Growth	Lpi Growth	Spi Growth	Loan Growth
Mean	0.034	0.052	0.000	0.016	0.004	0.035	0.068	0.077
Std. Dev.	0.043	0.036	0.017	0.193	0.014	0.074	0.202	0.054
Correlation Matrix								
	Inflation	Gensaki	Y-gap	Imp Growth	Ulo Growth	Lpi Growth	Spi Growth	Loan Growth
Inflation	1.000							
Gensaki	0.607	1.000						
Y-gap	−0.211	0.309	1.000					
Imp Growth	0.339	0.550	0.225	1.000				
Ulo Growth	0.492	0.198	−0.052	0.328	1.000			
Lpi Growth	0.185	0.777	0.591	0.345	−0.057	1.000		
Spi Growth	−0.069	−0.011	−0.286	−0.349	−0.176	0.081	1.000	
Loan Growth	0.489	0.823	0.310	0.279	−0.016	0.848	0.245	1.000

annualized quarterly data) are for the rates of growth of the share market and import price indices.

Table 7.6 shows that the highest correlation of inflation is with the Gensaki rate, but that it is positive rather than negative. This is another example of the well-known *price puzzle*, recently analyzed by Giordani (2001). This puzzle is also a common finding of linear vector autoregressive (VAR) models, which show that an increase in the interest rate has positive, rather than negative, effects on the price level in impulse-response analysis. Sims (1992) proposed that the cause of the prize puzzle may be unobservable contemporaneous supply shocks. The policymakers observe the shock and think it will have positive effects on inflation, so they raise the interest rates in anticipation of countering higher future inflation. Sims found that this puzzle disappears in U.S. data when we include a commodity price index in a more extensive VAR model.

Table 7.6 also shows that the second and third highest correlations of inflation are with unit labor costs and bank lending, followed by import price growth. The correlations of inflation with the share-price growth rate and the output gap are negative but insignificant.

Finally, what is most interesting from the information given in Table 7.6 is the very high correlation between the growth rate of bank lending and the growth rate of the land price index, not the growth rate of the share price index. It is not clear which way the causality runs: does the collapse of land prices lead to a fall in bank lending, or does the collapse of bank lending lead to a fall in land prices?

TABLE 7.7. Granger Test of Causality: LPI and Loan Growth

	Loan Growth Does Not Cause LPI Growth	LPI Growth Does Not Cause Loan Growth
F-Statistic	2.429	3.061
P-Value	0.053	0.020

In Japan, the story is different: banking credit and land prices show bidirectional causality or feedback. The collapse of land prices reduces bank lending, but the collapse of bank lending also leads to a fall in land prices. Table 7.7 gives the joint-F statistics and the corresponding P-values for a Granger test of causality. We see that the results are somewhat stronger for a causal effect from land prices to loan growth. However, the P-value for causality from loan growth to land price growth is only very slightly above 5%. These results indicate that both variables have independent influences and should be included as financial factors for assessing the behavior of inflation.

7.2.2 Model Specification

We use the same model specification for the Hong Kong deflation as in 7.1.2 with two exceptions: we do not use a price gap variable measuring convergence with mainland China, and we include both the domestic Gensaki interest rate and the rate of growth of bank lending as further explanatory variables for the evolution of inflation. As before, we forecast over a one-year horizon, and all rates of growth are measured as annual rates of growth, with $\Delta_h x_t = x_t - x_{t-h}$ and with $h = 4$.

7.2.3 In-Sample Performance

Figure 7.16 pictures the in-sample performance of the three models. The solid curve is for the error path of the linear model while similar dashed and dotted paths are the errors for alternative STRS and NNRS models. Both alternatives improve upon the performance of the linear model. Adding a bit of complexity greatly improves the statistical in-sample fit.

Table 7.8 gives the in-sample diagnostic statistics of the three models. We see that the STRS and NNRS models outperform the linear model, not only on the basis if goodness-of-fit measures, but also on specification tests. We can reject neither serial independence in the residuals nor the squared residuals for both alternative models. Similarly, we cannot reject normality in the residuals of both alternatives to the linear model. Finally, the Brock-Deckert-Scheinkman and Lee-White-Granger tests show there is very little or no evidence of neglected nonlinearity in the NNRS model.

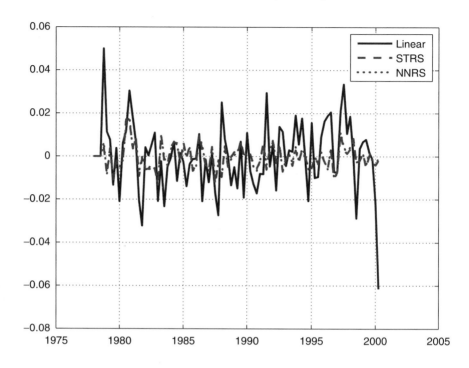

FIGURE 7.16. In-sample paths of estimation errors

The information from Table 7.8 gives strong support for abandoning a linear approach for understanding inflation/deflation dynamics in Japan.

7.2.4 Out-of-Sample Performance

Figure 7.17 gives the out-of-sample error paths of the three models. The solid curve is for the linear prediction errors, the dashed path is for the STRS prediction errors, and the dotted path is for the NNRS errors. We see that the NNRS models outperforms both the STRS and linear models. What is of interest, however, is that all three models generate negative prediction errors in 1997, the time of the onset of the Asian crisis. The models' negative errors, in which the errors represent differences between the actual and predicted outcomes, are indicators that the models do not incorporate the true depth of the deflationary process taking place in Japan.

Table 7.9 gives the out-of-sample test statistics of the three models. We see that the NNRS model has a much higher success ratio (in terms of percentage correct sign predictions of the dependent variable), and outperforms the linear model as well as the STRS model in terms of the root mean squared error statistic. The Diebold-Mariano statistics indicate that

TABLE 7.8. In-Sample Diagnostics of Alternative Models (Sample 1978–2002, Quarterly Data)

Diagnostics	Models		
	Linear	STRS	NNRS
SSE	0.023	0.003	0.003
RSQ	0.240	0.900	0.910
HQIF	−315.552	−466.018	−467.288
LB*	0.067	0.458	0.681
ML*	0.864	0.254	0.200
JB*	0.002	0.172	0.204
EN*	0.531	0.092	0.084
BDS*	0.012	0.210	0.119
LWG	484	56	3

*: prob value
Note:
SSE: Sum of squared errors
RSQ: R-squared
HIQF: Hannan-Quinn information criterion
LB: Ljung-Box Q statistic on residuals
ML: McLeod-Li Q statistic on squared residuals
JB: Jarque-Bera statistic on normality of residuals
EN: Engle-Ng test of symmetry of residuals
BDS: Brock-Deckert-Scheinkman test of nonlinearity
LWG: Lee-White-Granger test of nonlinearity

the NNRS prediction errors are statistically different from the linear model. However, the STRS prediction errors are not statistically different from either the linear or the NNRS model.

7.2.5   Interpretation of Results

The partial derivatives of the model for Japan, as well as the tests of significance based on bootstrapping methods, appear in Table 7.10. We see that the only significant variables determining future inflation are current inflation, the interest rate, and the rate of growth of the land price index. The output gap is almost, but not quite, significant. Unit labor costs and the Nikkei index are both insignificant and have the wrong sign.

The significant but wrong sign of the interest rate may be explained by the fact that the Bank of Japan is constrained by the zero lower bound of interest rates. They were lowering interest rates, but not enough during the period of deflation, so that real interest rates were in fact increasing. We see this in Figure 7.18.

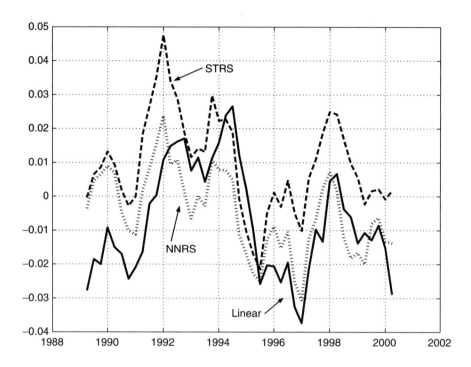

FIGURE 7.17. Out-of-sample prediction errors

TABLE 7.9. Out-of-Sample Forecasting Accuracy

Diagnostics	Models		
	Linear	STRS	NNRS
RMSQ	0.018	0.017	0.013
SR	0.511	0.489	0.644
Diebold-Mariano Test	Linear vs. STRS	Linear vs. NNRS	STRS vs. NNRS
DM-1*	0.276	0.011	0.233
DM-2*	0.304	0.016	0.271
DM-3*	0.310	0.007	0.285
DM-4*	0.306	0.001	0.289
DM-5*	0.301	0.001	0.288

*: prob value
RMSQ: Root mean squared error
SR: Success ratio on sign correct sign predictions
DM: Diebold-Mariano test
(correct for autocorrelation, lags 1–5)

TABLE 7.10. Partial Derivatives of NNRS Model

Period	Arguments							
	Inflation	Interest Rate	Import Price	Lending Growth	Nikkei Index	Land Price Index	Output Gap	Unit Labor Costs
Mean	0.182	0.212	0.113	0.025	−0.088	0.122	0.015	−0.075
1978	0.190	0.217	0.123	0.039	−0.089	0.112	0.019	−0.092
1995	0.183	0.212	0.114	0.026	−0.088	0.121	0.015	−0.077
2002	0.181	0.211	0.112	0.023	−0.087	0.124	0.015	−0.074

Statistical Significance of Estimates

Period	Arguments							
	Inflation	Interest Rate	Import Price	Lending Growth	Nikkei Index	Land Price Index	Output Gap	Unit Labor Costs
Mean	0.000	0.000	0.859	0.935	0.356	0.000	0.149	1.000
1978	0.000	0.000	0.819	0.933	0.288	0.000	0.164	1.000
1995	0.000	0.000	0.840	0.931	0.299	0.000	0.164	1.000
2002	0.000	0.000	0.838	0.935	0.293	0.000	0.149	1.000

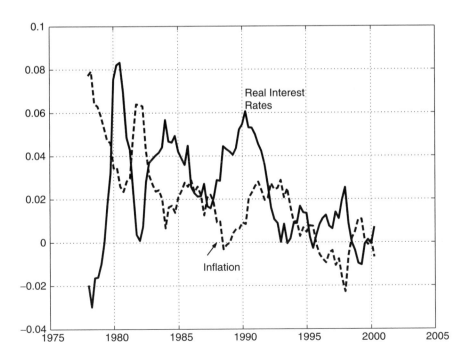

FIGURE 7.18. Real interest rates and inflation in Japan

The fact that the land price index is significant while the Nikkei index is not can be better understood by looking at Figure 7.14. The rate of growth has shown a smooth steady decline, more in tandem with the inflation process than with the much more volatile Nikkei index.

Table 7.11 gives the corresponding sets of partial derivatives and tests of significance from the STRS model. The only difference we see from the NNRS model is that the output gap variable is also significant.

Figure 7.19 pictures the evolution of inflation and the transition neurons of the two models. As in the case of Hong Kong, the STRS transition neuron gives more information, showing that the likelihood of remaining in the inflation state is steadily decreasing as inflation switches to deflation after 1995. The NNRS model's transition neuron shows little or no action, remaining close to 0.5. The result indicates that the NNRS model outperforms the linear and STRS model not by picking up a regime change *per se* but rather by approximating nonlinear processes in the overall inflation process.

The fact that bank lending does not appear as a significant determinant of inflation (while output gap does — at least in the STRS model) does not mean that bank lending is not important. Table 7.12 pictures the results of a Granger causality test between the output gap and the rate of growth of bank lending in Japan. We see strong evidence, at the 5% level

TABLE 7.11. Partial Derivatives of STRS Model

Period	Arguments							
	Inflation	Interest Rate	Import Price	Lending Growth	Nikkei Index	Land Price Index	Output Gap	Unit Labor Costs
Mean	0.149	0.182	0.054	−0.094	−0.032	0.208	0.028	−0.079
1978	0.138	0.163	0.055	−0.096	−0.032	0.232	0.030	−0.080
1995	0.138	0.163	0.055	−0.096	−0.032	0.232	0.030	−0.080
2002	0.133	0.156	0.056	−0.096	−0.032	0.242	0.030	−0.080

Statistical Significance of Estimates

Period	Arguments							
	Inflation	Interest Rate	Import Price	Lending Growth	Nikkei Index	Land Price Index	Output Gap	Unit Labor Costs
Mean	0.006	0.000	0.695	1.000	0.398	0.000	0.095	1.000
1978	0.006	0.000	0.695	1.000	0.398	0.000	0.095	1.000
1995	0.006	0.000	0.615	1.000	0.394	0.000	0.088	0.863
2002	0.002	0.000	0.947	1.000	0.739	0.000	0.114	1.000

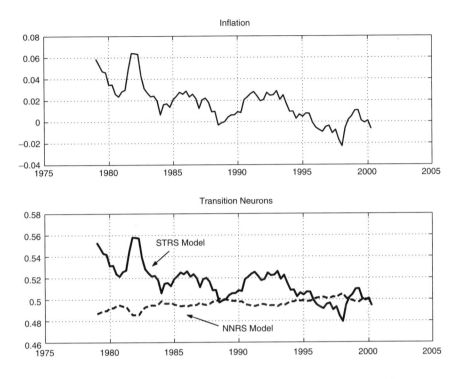

FIGURE 7.19. Regime transitions in STRS and NNRS models

TABLE 7.12. Ganger Test of Causality: Loan Growth and the Output Gap

	Hypothesis	
	Loan Growth Does Not Cause the Output Gap	Output Gap Does Not Cause Loan Growth
F-Statistic	2.5	2.4
P-Value	0.049	0.053

of significance, that the rate of growth of bank loans is a causal factor for changes in the output gap. There is also evidence of reverse causality, from the output gap to the rate of growth of bank lending, to be sure. These results indicate that a reversal in bank lending will improve the output gap, and such an improvement will call forth more bank lending, leading, in turn, in a virtuous cycle, to further output-gap improvement and an escape from the deflationary trap in Japan.

7.3 Conclusion

The chapter illustrates how neural network regime switching models help explain the evolution of inflation and deflation in Japan and Hong Kong. The results for Hong Kong indicate that external prices and residential property prices are the most important factors underlying inflationary dynamics, whereas for Japan, interest rates and excess demand (proxied by the output gap) appear to be more important. These results are consistent with well-known stylized facts about both economies. Hong Kong is a much smaller and more highly open economy than Japan, so that the evolution of international prices and nontraded prices (proxied by residential property prices) would be the driving forces behind inflation. For Japan, a larger and less open economy, we would expect policy variables and excess demand to be more important factors for inflation.

Clearly, there are a large number of alternative nonlinear as well as neural network specifications for approximating the inflation processes of different countries. We used a regime switching approach since both Hong Kong and Japan have indeed moved from inflationary to deflationary regimes. But for most countries, the change in regime may be much different, such as an implicit or explicit switch to inflation-targets for monetary policy. These types of regime switches cannot be captured as easily as the switch from inflation to deflation.

Since inflation is of such central importance for both policymakers and decision makers in business, finance, and households, it is surprising that more work using neural networks has not been forthcoming. Chen, Racine, and Swanson (2001) have used a ridgelet neural network for forecasting inflation in the United States. McNelis and McAdam (2004) used a thick model approach (combining forecasts of different types of neural nets) for both the Euro Zone and the United States. Both of these papers show the improved forecasting performance from neural network methods. Hopefully, more work will follow.

7.3.1 MATLAB Program Notes

The same programs used in the previous chapter were used for the inflation/deflation studies. The data are given in *honkonginflation_may2004_run8.mat* and *japdata_may2004_run3.mat* for Hong Kong and Japan.

7.3.2 Suggested Exercises

The reader is invited to use data from other countries to see how well the results from Japan or Hong Kong carry over to countries that did not experience deflation as well as inflation. However, the threshold would have to be changed from zero to a very low positive inflation level. What would be of interest is the role of residential property prices as a key variable driving inflation.

8

Classification: Credit Card Default and Bank Failures

This chapter examines how well neural network methods compare with more traditional methods based on discriminant analysis, as well as nonlinear logit, probit, and Weibull methods, spelled out in Chapter 2, Section 7. We examine two cases, one for classification of credit card default using German data, and the other for banking intervention or closure, using data from Texas in the 1980s. Both of these data sets and the results we show are solely meant to be examples of neural network performance relative to more traditional econometric methods. There is no claim to give new insight into credit card risk assessment or early warning signals for a banking problem.

We see in both of the examples that classification problems involve the use of numerical indicators for qualitative characteristics such as gender, marital status, home ownership, or membership in the Federal Reserve System. In this case, we are using *crisp logic* or *crisp sets*: a person is either in one group or another. However, a related method for classification involves *fuzzy sets* or *fuzzy logic*, in which a person may be partially in one category or another (as in health studies, for example, one may be partially overweight: partly in one set of "overweight" and partly in the other set of "normal" weight). Much of the related artificial intelligence "neuro-fuzzy" literature related to neural nets and fuzzy logic has focused on deriving rules for making decisions, based on the outcome of classification schemes. In this chapter, however, we will simply focus on the neural network approach with respect to the traditional linear discriminant analysis and the nonlinear logit, probit, and Weibull methods.

When working with any nonlinear function, however, we should never underestimate the difficulties of obtaining optima, even with simple probit or Weibull models used for classification. The logit model, of course, is a special case of the neural network, since a neural network with one logsigmoid neuron reduces to the logit model. But the same tools we examined in previous chapters — particularly hybridization or coupling the genetic algorithm with quasi-Newton gradient methods — come in very handy. Classification problems involving nonlinear functions have all of the same problems as other models, especially when we work with a large number of variables.

8.1   Credit Card Risk

For examining credit card risk, we make use of a data set used by Baesens, Setiono, Mues, and Vanthienen (2003), on German credit card default rates. The data set we use for classification of default/no default for German credit cards consists of 1000 observations.

8.1.1   The Data

Table 8.1 lists the twenty arguments, a mix of categorical and continuous variables. Table 8.1 also gives the maximum, minimum, and median values of each of the variables. The dependent variable y takes on a value of 0 if there is no default and a value of 1 if there is a default. There are 300 cases of defaults in this sample, with $y = 1$. As we can see in the mix of variables, there is considerable discretion about how to categorize the information.

8.1.2   In-Sample Performance

The in-sample performance of the five methods appears in Table 8.2. This table pictures both the likelihood functions for the four nonlinear alternatives to the discriminant analysis and the error percentages of all five methods. There are two types of errors, as taught from statistical decision theory. False positives take place when we incorrectly label the dependent variables as 1, with $\widehat{y} = 1$ when $y = 0$. Similarly, false negatives occur when we have $\widehat{y} = 0$ when $y = 1$. The overall error ratio in Table 8.2 is simply a weighted average of the two error percentages, with the weight set at .5.

In the real world, of course, decision makers attach differing weights to the two types of errors. A false positive means that a credit agency or bank incorrectly denies a credit card to a potentially good customer and thus loses revenue from a reliable transaction. A false negative is more serious: it means extending credit to a potentially unreliable customer, and thus the bank assumes much higher default risk.

TABLE 8.1. Attributes for German Credit Data Set

Variable	Definition	Type/Explanation	Max	Min	Median
1	Checking account	Categorical, 0 to 3	3	0	1
2	Term	Continuous	72	4	18
3	Credit history	Categorical, 0 to 4, from no history to delays	4	0	2
4	Purpose	Categorical, 0 to 9, based on type of purchase	10	0	2
5	Credit amount	Continuous	18424	250	2319.5
6	Savings account	Categorical, 0 to 4, lower to higher to unknown	4	0	1
7	Yrs in present employment	Categorical, 0 to 4, 1 unemployment, to longer years	4	0	2
8	Installment rate	Continuous	4	1	3
9	Personal status and gender	Categorical, 0 to 5, 1 male, divorced, 5 female, single	3	0	2
10	Other parties	Categorical, 0 to 2, none, 2 co-applicant, 3 guarantor	2	0	0
11	Yrs in present residence	Continuous	4	1	3
12	Property type	Categorical, 0 to 3, 0 real estate, 3 no property or unknown	3	0	2
13	Age	Continuous	75	19	33
14	Other installment plans	Categorical, 0 to 2, 0 bank, 1 stores, 2 none	2	0	0
15	Housing status	Categorical, 0 to 2, 0 rent, 1 own, 2 for free	2	0	2
16	Number of existing credits	Continuous	4	1	1
17	Job status	Categorical, 0 to 3, unemployed, 3 management	3	0	2
18	Number of dependents	Continuous	2	1	1
19	Telephone	Categorical, 0 to 1, 0 none, 1 yes, under customer name	1	0	0
20	Foreign worker	Categorical, 0 to 1, 0 yes, 1 no	1	0	0

TABLE 8.2. Error Percentages

Method	Likelihood Fn.	False Positives	False Negatives	Weighted Average
Discriminant analysis	na	0.207	0.091	0.149
Neural network	519.8657	0.062	0.197	0.1295
Logit	519.8657	0.062	0.197	0.1295
Probit	519.1029	0.062	0.199	0.1305
Weibull	516.507	0.072	0.189	0.1305

The neural network alternative to the logit, probit, and Weibull methods is a network with three neurons. In this case, it is quite similar to a logit model, and in fact the error percentages and likelihood functions are identical. We see in Table 8.2 a familiar trade-off. Discriminant analysis has fewer false negatives, but a much higher percentage (by more than a factor of three) of false positives.

8.1.3   Out-of-Sample Performance

To evaluate the out-of-sample forecasting accuracy of the alternative models, we used the 0.632 bootstrap method described in Section 4.2.8. To summarize this method, we simply took 1000 random draws of data from the original sample, with replacement, to do an estimation, and thus used the excluded data from the original sample to evaluate the out-of-sample forecast performance. We measured the out-of-sample forecast performance by the error percentages of false positives or false negatives. We repeated this process 100 times and examined the mean and distribution of the error-percentages of the alternative models.

Table 8.3 gives the mean error percentages for each method, based on the bootstrap experiments. We see that the neural network and logit models give identical performance, in terms of out-of-sample accuracy. We also see that discriminant analysis and the probit and Weibull methods are almost mirror images of each other. Whereas discriminant analysis is perfectly accurate in terms of false positives, it is extremely imprecise (with an error rate of more than 75%) in terms of false negatives, while probit and Weibull are quite accurate in terms of false negatives, but highly imprecise in terms of false positives. The better choice would be to use logit or the neural network method.

The fact that the network model does not outperform the logit model should not be a major cause for concern. The logit model is a neural net model with one neuron. The network we use is a model with three neurons. Comparing logit and neural network models is really a comparison of two alternative neural network specifications, one with one neuron and

TABLE 8.3. Out-of-Sample Forecasting: 100 Draws Mean Error Percentages (0.632 Bootstarp)

Method	False Positives	False Negatives	Weighted Average
Discriminant analysis	0.000	0.763	0.382
Neural network	0.095	0.196	0.146
Logit	0.095	0.196	0.146
Probit	0.702	0.003	0.352
Weibull	0.708	0.000	0.354

another with three. What is surprising is that the introduction of the additional two neurons in the network does not cause a deterioration of the out-of-sample performance of the model. By adding the two additional neurons we are not overfitting the data or introducing nuisance parameters which cause a decline in the predictive performance of the model. What the results indicate is that the class of parsimoniously specified neural network models greatly outperforms discriminant analysis, probit, and Weibull specifications.

Figure 8.1 pictures the distribution of the weighted average (of false positives and negatives) for the two models over the 100 bootstrap experiments. We see that they are identical.

8.1.4 Interpretation of Results

Table 8.4 gives information on the partial derivatives of the models as well as the corresponding marginal significance or P-values of these estimates, based on the bootstrap distributions. We see that the estimates of the network and logit models are for all practical purposes identical. The probit model results do not differ by much, whereas the Weibull estimates differ by a bit more, but not by a large factor.

Many studies using classification methods are not interested in the partial derivatives, since interpretation of specific categorical variables is not as straightforward as continuous variables. However, the bootstrapped P-values show that credit amount, property type, job status, and number of dependents are not significant. Some results are consistent with expectations: the greater the number of years in present employment, the lower the risk of a default. Similarly for age, telephone, other parties, or status as a foreign worker: older persons, who have telephones in their own name, have partners in their account, and are not foreign are less likely to default, We also see that having a higher installment rate or multiple installment plans is more likely to lead to default.

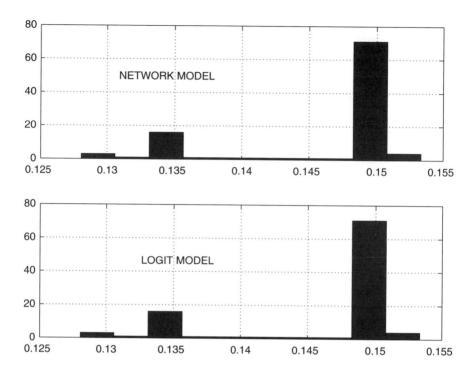

FIGURE 8.1. Distribution of 0.632 bootstrap out-of-sample error percentages

While all three models give broadly consistent interpretations, this should be reassuring rather than a cause of concern. These results indicate that using two methods, logit and neural net, one as a check on the other, may be sufficient for both accuracy and understanding.

8.2   Banking Intervention

Banking intervention, the need to close or to put a private bank under state management, more extensive supervision, or to impose a change of management, is, unfortunately, common enough both in developing and in mature industrialized countries. We use the same binary or classification methods to examine how well key characteristics of banks may serve as early warning signals for a crisis or intervention of a particular bank.

8.2.1   The Data

Table 8.5 gives information about the dependent variables as well as explanatory variables we use for our banking study. The data were obtained

TABLE 8.4.

Variable	Definition	Partial Derivatives*				Prob Values**			
		Network	Logit	Probit	Weibull	Network	Logit	Probit	Weibull
1	Checking account	0.074	0.074	0.076	0.083	0.000	0.000	0.000	0.000
2	Term	0.004	0.004	0.004	0.004	0.000	0.000	0.000	0.000
3	Credit history	−0.078	−0.078	−0.077	−0.076	0.000	0.000	0.000	0.000
4	Propose	−0.007	−0.007	−0.007	−0.007	0.000	0.000	0.000	0.000
5	Credit amount	0.000	0.000	0.000	0.000	0.150	0.150	0.152	0.000
6	Savings account	−0.008	−0.008	−0.009	−0.010	0.020	0.020	0.020	0.050
7	Yrs in present employment	−0.032	−0.032	−0.031	−0.030	0.000	0.000	0.000	0.000
8	Installment rate	0.053	0.053	0.053	0.049	0.000	0.000	0.000	0.000
9	Personal status and gender	−0.052	−0.052	−0.051	−0.047	0.000	0.000	0.000	0.000
10	Other parties	−0.029	−0.029	−0.026	−0.020	0.010	0.010	0.020	0.040
11	Yrs in present residence	0.008	0.008	0.008	0.004	0.050	0.050	0.040	0.060
12	Property type	−0.002	−0.002	−0.000	0.003	0.260	0.260	0.263	0.300
13	Age	−0.003	−0.003	−0.003	−0.002	0.000	0.000	0.000	0.010
14	Other installment plans	0.057	0.057	0.062	0.073	0.000	0.000	0.000	0.000
15	Housing status	−0.047	−0.047	−0.050	−0.051	0.000	0.000	0.000	0.000
16	Number of existing credits	0.057	0.057	0.055	0.053	0.000	0.000	0.000	0.000
17	Job status	0.003	0.003	0.006	0.012	0.920	0.920	0.232	0.210
18	Number of dependents	0.032	0.032	0.030	0.022	0.710	0.710	0.717	0.030
19	Telephone	−0.064	−0.064	−0.065	−0.067	0.000	0.000	0.000	0.000
20	Foreign worker	−0.165	−0.165	−0.153	−0.135	0.000	0.000	0.000	0.000

*: Derivatives calculated as finite differences
**: Prob values calculated from bootstrap distributions

from the Federal Reserve Bank of Dallas using banking records from the last two decades. The total percentage of banks that required intervention, either by state or federal authorities, was 16.7. We use 12 variables as arguments. The capital-asset ratio, of course, is the key component of the well-known Basel accord for international banking standards.

While the negative number for the minimum of the capital-asset ratio may seem surprising, the data set includes both sound and unsound banks. When we remove the observations having negative capital-asset ratios, the distribution of this variable shows that the ratio is between 5 and 10% for most of the banks in the sample. The distribution appears in Figure 8.2.

8.2.2 In-Sample Performance

Table 8.6 gives information about the in-sample performance of the alternative models.

TABLE 8.5. Texas Banking Data

		Max	Min	Median
1	Charter	1	0	0
2	Federal Reserve	1	0	1
3	Capital/asset %	30.9	−77.71	7.89
4	Agricultural loan/total loan ratio	0.822371	0	0.013794
5	Consumer loan/total loan ratio	0.982775	0	0.173709
6	Credit card loan/total loan ratio	0.322974	0	0
7	Installment loan/total loan ratio	0.903586	0	0.123526
8	Nonperforming loan/total loan - %	35.99	0	1.91
9	Return on assets - %	10.06	−36.05	0.97
10	Interest margin - %	10.53	−2.27	3.73
11	Liquid assets/total assets - %	96.54	3.55	52.35
12	U.S. total loans/U.S. gdp ratio	2.21	0.99	1.27

Dependent Variables: Bank closing or intervention
No observations: 12,605
% of Interventions/closings: 16.7

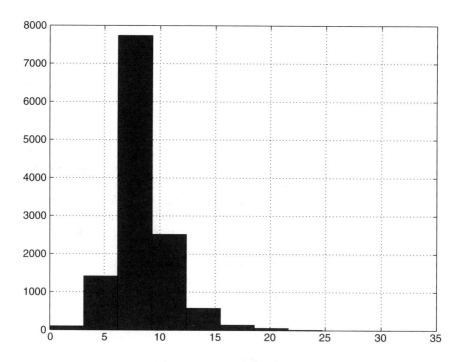

FIGURE 8.2. Distribution of capital-asset ratio (%)

TABLE 8.6. Error Percentages

Method	Likelihood Fn.	False Positives	False Negatives	Weighted Average
Discriminant analysis	na	0.205	0.038	0.122
Neural network	65535	0.032	0.117	0.075
Logit	65535	0.092	0.092	0.092
Probit	4041.349	0.026	0.122	0.074
Weibull	65535	0.040	0.111	0.075

TABLE 8.7. Out-of-Sample Forecasting: 40 Draws Mean Error Percentages (0.632 Bootstarp)

Method	False Positives	False Negatives	Weighted Average
Discriminant analysis	0.000	0.802	0.401
Neural network	0.035	0.111	0.073
Logit	0.035	0.089	0.107
Probit	0.829	0.000	0.415
Weibull	0.638	0.041	0.340

Similar to the example with the credit card data, we see that discriminant analysis gives more false positives than the competing nonlinear methods. In turn, the nonlinear methods give more false negatives than the linear discriminant method. For overall performance, the network, probit, and Weibull methods are about the same, in terms of the weighted average error score. We can conclude that the network model, specified with three neurons, performs about as well as the most accurate method, for in-sample estimation.

8.2.3 Out-of-Sample Performance

Table 8.7 gives the mean error percentages, based on the 0.632 bootstrap method. The ratios are the averages over 40 draws, by the bootstrap method. We see that discriminant analysis has a perfect score, zero percent, on false positives, but has a score of over 80% on false negatives. The overall best performance in this experiment is by the neural network, with a 7.3% weighted average error score. The logit model is next, with a 10% weighted average score. As in the previous example the neural network family outperforms the other methods in terms of out-of-sample accuracy.

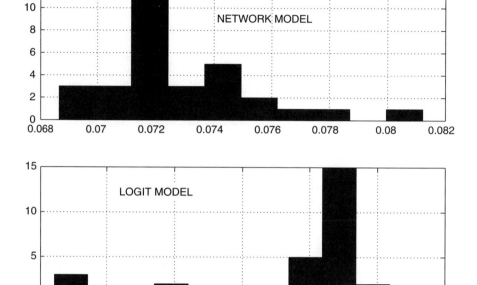

FIGURE 8.3. Distribution of 0.632 bootstrap: out-of-sample error percentages

Figure 8.3 pictures the distribution of the out-of-sample weighted average error scores of the network and logit models. While the average of the logit model is about 10%, we see in this figure that the center of the distribution, for most of the data, is between 11 and 12%, whereas the corresponding center for the network model is between 7.2 and 7.3%. The network model's performance clearly indicates that it should be the preferred method for predicting individual banking crises.

8.2.4   Interpretation of Results

Table 8.8 gives the partial derivatives as well as the corresponding P-values (based on bootstrapped distributions). Unlike the previous example, we do not have the same broad consistency about the signs or significance of the key variables. However, what does emerge is the central importance of the capital asset ratio as an indicator of banking vulnerability. The higher this ratio, the lower the likelihood of banking fragility. Three of the four models (network, logit, and probit) indicate that this variable is significant, and the magnitude of the derivatives (calculated by finite differences) is the same.

TABLE 8.8.

No.	Definition	Partial Derivatives*				Prob Values**			
		Network	Logit	Probit	Weibull	Network	Logit	Probit	Weibull
1	Charter	0.000	0.000	−0.109	−0.109	0.767	0.833	0.267	0.533
2	Federal Reserve	0.082	0.064	0.031	0.031	0.100	0.167	0.000	0.400
3	Capital/asset %	−0.051	−0.036	−0.053	−0.053	0.000	0.000	0.000	0.367
4	Agricultural loan/ total loan ratio	0.257	0.065	−0.020	−0.020	0.133	0.200	0.000	0.600
5	Consumer loan/ total loan ratio	0.397	0.088	0.094	0.094	0.300	0.767	0.000	0.433
6	Credit card loan/ total loan ratio	1.049	−1.163	−0.012	−0.012	0.700	0.233	0.000	0.567
7	Installment loan/ total loan ratio	−0.137	0.187	−0.115	−0.115	0.967	0.233	0.000	0.600
8	Nonperforming loan/total loan - %	0.004	0.001	0.010	0.010	0.167	0.167	0.067	0.533
9	Return on assets - %	−0.042	−0.025	−0.032	−0.032	0.067	0.133	0.000	0.367
10	Interest margin - %	0.013	−0.029	0.018	0.018	0.967	0.933	1.000	0.567
11	Liquid assets/ total assets - %	0.001	0.002	0.001	0.001	0.067	0.667	0.000	0.533
12	U.S. total loans/ U.S. gdp ratio	0.149	0.196	0.118	0.118	0.000	0.033	0.000	0.333

*: Derivatives calculated as finite differences
**: Prob values calculated from bootstrap distributions

The same three models also indicate that the aggregate U.S. total loan to total GDP ratio is also a significant determinant of an individual bank's fragility. Thus, both aggregate macro conditions and individual bank characteristics matter, as informative signals for banking problems. Finally, the network model (as well as the probit) show that return on assets is also significant as an indicator, with a higher return, as expected, lowering the likelihood of banking fragility.

8.3   Conclusion

In this chapter we examined two data sets, one on credit card default rates, and the other on banking failures or fragilities requiring government intervention. We found that neural nets either perform as well as or better than the best nonlinear alternative, from the set of logit, probit, or Weibull models, for classification. The hybrid evolutionary genetic algorithm and classical gradient-descent methods were used to obtain the parameter estimates for all of the nonlinear models. So we were not handicapping one or another model with a less efficient estimation process. On the contrary, we did the best to find, as closely as possible, the global optima when maximizing the likelihood functions.

There are clearly many interesting examples to study with this methodology. The work on early warning signals for currency crises would be amenable to this methodology. Similarly, further work comparing neural networks to standard models can be done on classification problems involving more than two categories, or on discrete ordered multinomial problems, such as student evaluation rankings of professors on a scale of one through five [see Evans and McNelis (2000)].

The methods in this chapter could be extended into more elaborate networks in which the predictions of different models, such as discriminant, logit, probit, and Weibull, are fed in as inputs to a complex neural network. Similarly, forecasting can be done in a thick modeling or bagging approach: all of the models can be used, and a mean or trimmed mean can be the forecast from a wide set of models, including a variety of neural nets specified with different numbers of neurons in the hidden layer. But in this chapter we wanted to keep the "race" simple, so we leave the development of more elaborate networks for further exploration.

8.3.1   MATLAB Program Notes

The programs for these two country experiences are *germandefault_prog.m* for German credit card default rates, and *texasfinance_prog.m* for the Texas bank failures. The data are given in *germandefault_run4.mat* and *texasfinance_run9.mat*.

8.3.2   Suggested Exercises

An interesting sensitivity analysis would be to reduce the number of explanatory variables used in this chapter's examples to smaller sets of regressors to see if the same variables remain significant in the modified models.

9
Dimensionality Reduction and Implied Volatility Forecasting

In this chapter we apply the methodologies of linear and nonlinear principal component dimensionality reduction to observed volatilities on Hong Kong and United States swap options of differing maturities, of one to ten years, to see if these methods help us to find the underlying volatility signal from the market. The methods are presented in Section 2.6.

Obtaining an accurate measure of the market volatility, when in fact there are many different market volatility measures or alternative nonmarket measures of volatility to choose from, is a major task for effective option pricing and related hedging activities. A major focus in financial market research today is volatility, rather than return, forecasting. Volatilities, as proxies of risk, are asymmetric and perhaps nonlinear processes, at the very least, to the extent that they are bounded by zero from below. So nonlinear approximation methods such as neural networks may have a payoff when we examine such processes.

We compare and contrast the implied volatility measures for Hong Kong and the United States, since we expect both of these to have similar features, due to the currency peg of the Hong Kong dollar to the U.S. dollar. But there may also be some differences, since Hong Kong was more vulnerable to the Asian financial crisis which began in 1997, and also had the SARS crisis in 2003. We discuss both of these experiences in turn, and apply the linear and nonlinear dimensionality reduction methods for in-sample as well as for out-of-sample performance.

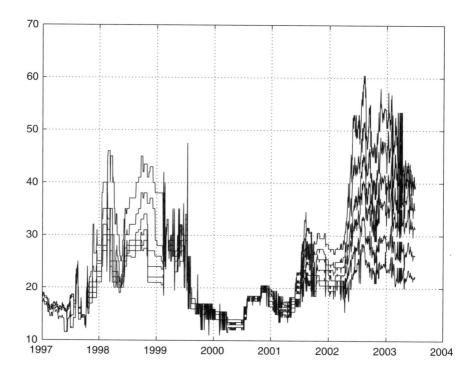

FIGURE 9.1. Hong Kong implied volatility measures, maturity 2, 3, 4, 5, 7, 10 years

9.1 Hong Kong

9.1.1 The Data

The implied volatility measures, for daily data from January 1997 till July 2003, obtained from Reuters, appear in Figure 9.1. We see the sharp upturn in the measures with the onset of the Asian crisis in late 1997. There are two other spikes: one around the third quarter of 2001, and another after the start of 2002. Both of these jumps, no doubt, reflect uncertainty in the world economy in the wake of the September 11 terrorist attacks and the start of the war in Afghanistan. The continuing volatility in 2003 may also be explained by the SARS epidemic in Hong Kong and East Asia.

Table 9.1 gives a statistical summary of the data appearing in Figure 9.1. There are a number of interesting features coming from this summary. One is that both the mean of the implied volatilities, as well as the standard

TABLE 9.1. Hong Kong Implied Volatility Estimates; Daily Data: Jan. 1997–July 2003

Statistic	Maturity in Years					
	2	3	4	5	7	10
Mean	28.581	26.192	24.286	22.951	21.295	19.936
Median	27.500	25.000	23.500	22.300	21.000	20.000
Std. Dev.	12.906	10.183	8.123	6.719	5.238	4.303
Coeff. Var	0.4516	0.3888	0.33448	0.2927	0.246	0.216
Skewness	0.487	0.590	0.582	0.536	0.404	0.584
Kurtosis	2.064	2.235	2.302	2.242	2.338	3.553
Max	60.500	53.300	47.250	47.500	47.500	47.500
Min	11.000	12.000	12.250	12.750	12.000	11.000

deviation of the implied volatility measures, or volatility of the volatilities, decline as the maturity increases. Related to this feature is that the range, or difference between maximum and minimum values, is greatest for the short maturity of two years. The extent of the variability decline in the data can best be captured by the coefficient of variation, defined as the ratio of the standard deviation to the mean. We see that this measure declines by more than 50% as we move from two-year to ten-year maturities. Finally, there is no excess kurtosis in these measures, whereas rates of return typically have this property.

9.1.2   In-Sample Performance

Figure 9.2 pictures the evolution of the two principal component measures. The solid curve comes from the linear method. The broken curve comes from an auto-associative map or neural network. We estimate the network with five encoding neurons and five decoding neurons. For ease of comparison, we scaled each series between zero and one. What is most interesting about Figure 9.2 is how similar both curves are. The linear principal component shows a big spike in mid-1999, but the overall volatility of the nonlinear principal component is slightly greater. The standard deviations of the linear and nonlinear components are, respectively, .233 and .272, where their respective coefficients of variation are .674 and .724.

How well do these components explain the variation of the data, for the full sample? Table 9.2 gives simple goodness-of-fit R^2 measures for each of the maturities. We see that the nonlinear principal component better fits the more volatile 2-year maturity, whereas the linear component fits much, much better at 5, 7, and 10-year maturities.

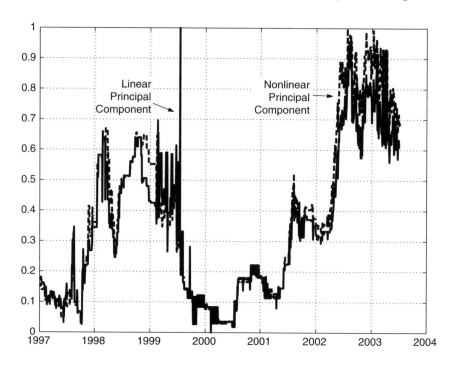

FIGURE 9.2. Hong Kong linear and nonlinear principal component measures

TABLE 9.2. Hong Kong Implied Volatility Estimates Goodness of Fit: Linear and Nonlinear Components, Multiple Correlation Coefficient

	Maturity in Years					
	2	3	4	5	7	10
Linear	0.965	0.986	0.990	0.981	0.923	0.751
Nonlinear	0.988	0.978	0.947	0.913	0.829	0.698

9.1.3   Out-of-Sample Performance

To evaluate the out-of-sample performance of each of the models, we did a recursive estimation of the principal components. First, we took the first 80% of the data, estimated the principal component coefficients and nonlinear functions for extracting one component, brought in the next observation, and applied these coefficients and functions for estimating the new principal component. We used this new forecast principal component

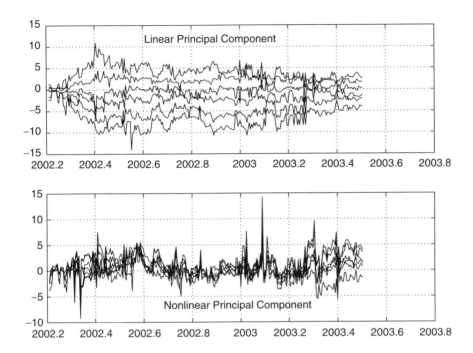

FIGURE 9.3. Hong Kong recursive out-of-sample principal component prediction errors

to explain the six observed volatilities at that observation. We then continued this process, adding in one observation each period, updating the sample, and re-estimating the coefficients and nonlinear functions, until the end of the data set.

The forecast errors of the recursively updated principal components appear in Figure 9.3. It is clear that the errors of the nonlinear principal component forecasting model are generally smaller than those of the linear principal component model. The most noticeable jump in the nonlinear forecast errors takes place in early 2003, at the time of the SARS epidemic in Hong Kong.

Are the forecast errors significantly different from each other? Table 9.3 gives the root mean squared error statistics as well as Diebold-Mariano tests of significance for these forecast errors, for each of the volatility measures. The results show that the nonlinear principal components do significantly better than the linear principal components at maturities of 2, 3, 7, and 10 years.

TABLE 9.3. Hong Kong Implied Volatility Estimates: Out-of-Sample Prediction Performance, Root Mean Squared Error

	Maturity in Years					
	2	3	4	5	7	10
Linear	4.195	2.384	1.270	2.111	4.860	7.309
Nonlinear	1.873	1.986	2.598	2.479	1.718	1.636
	Diebold-Mariano Tests[*] Maturity in Years					
	2	3	4	5	7	10
DM-0	0.000	0.000	1.000	0.762	0.000	0.000
DM-1	0.000	0.000	1.000	0.717	0.000	0.000
DM-2	0.000	0.000	1.000	0.694	0.000	0.000
DM-3	0.000	0.000	1.000	0.678	0.000	0.000
DM-4	0.000	0.000	1.000	0.666	0.000	0.000

Note: [*]P-values
DM-0 to DM-4: tests at autocorrelations 0 to 4.

9.2 United States

9.2.1 The Data

Figure 9.4 pictures the implied volatility measures for the same time period as the Hong Kong data, for the same maturities. While the general pattern is similar, we see that there is less volatility in the volatility measures in 1997 and 1998. There is a spike in the data in late 1998. The jump in volatility in later 2001 is of course related to the September 11 terrorist attacks, and the further increased volatility beginning in 2002 is related to the start of hostilities in the Gulf region and Afghanistan.

The statistical summary of these data appear in Table 9.4. The overall volatility indices of the volatilities, measured by the standard deviations and the coefficients of variation, are actually somewhat higher for the United States than for Hong Kong. But otherwise, we observe the same general properties that we see in the Hong Kong data set.

9.2.2 In-Sample Performance

Figure 9.5 pictures the linear and nonlinear principal components for the U.S. data. As in the case of Hong Kong, the volatility of the nonlinear principal component is greater than that of the linear principal component.

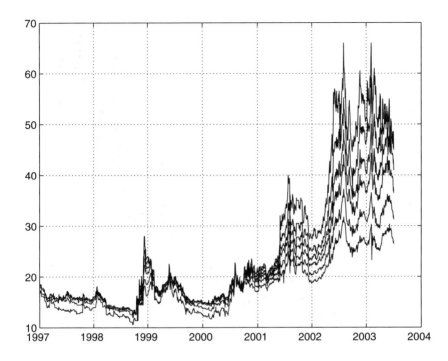

FIGURE 9.4. U.S. implied volatility measures, maturities 2, 3, 4, 5, 7, 10 years

TABLE 9.4. U.S. Implied Volatility Estimates, Daily Data: Jan. 1997–July 2003

Statistic	Maturity in Years					
	2	3	4	5	7	10
Mean	24.746	23.864	22.799	21.866	20.360	18.891
Median	17.870	18.500	18.900	19.000	18.500	17.600
Std. Dev.	14.621	11.925	9.758	8.137	6.106	4.506
Coeff. Var	0.591	0.500	0.428	0.372	0.300	0.239
Skewness	1.122	1.214	1.223	1.191	1.092	0.952
Kurtosis	2.867	3.114	3.186	3.156	3.023	2.831
Max	66.000	59.000	50.000	44.300	37.200	31.700
Min	10.600	12.000	12.500	12.875	12.750	12.600

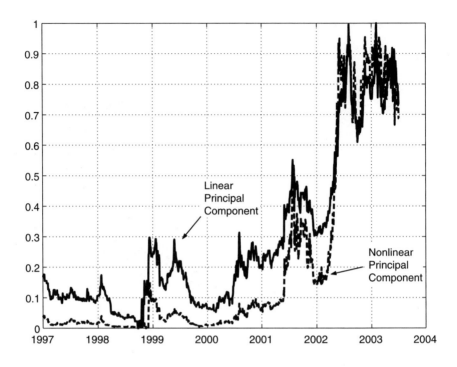

FIGURE 9.5. U.S. linear and nonlinear principal component measures

TABLE 9.5. U.S. Implied Volatility Estimates Goodness of Fit: Linear and Nonlinear Components Multiple Correlation Coefficient

	Maturity in Years					
	2	3	4	5	7	10
Linear	0.983	0.995	0.997	0.998	0.994	0.978
Nonlinear	0.995	0.989	0.984	0.982	0.977	0.969

The goodness-of-fit R^2 measures appear in Table 9.5. We see that there is not as great a drop-off in the explanatory power of the two components, as in the case of Hong Kong, as we move up the maturity scale.

9.2.3   Out-of-Sample Performance

The recursively estimated out-of-sample prediction errors of the two components appear in Figure 9.6. As in the case of Hong Kong, the prediction errors of the nonlinear component appear to be more tightly clustered.

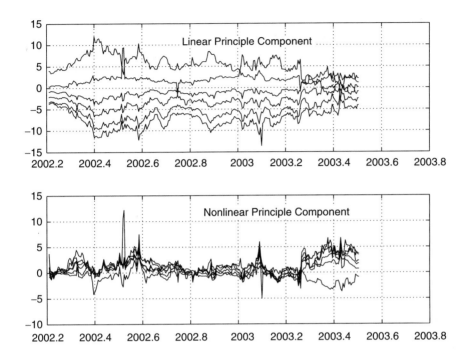

FIGURE 9.6. U.S. recursive out-of-sample principal component prediction errors

There are noticeable jumps in the nonlinear prediction errors in mid-2002 and in 2003 at the end of the sample.

The root mean squared error statistics as well as the Diebold-Mariano tests of significance appear in Table 9.5. For the United States, the nonlinear component outperforms the linear component for all maturities except for four years.[1]

9.3   Conclusion

In this chapter we examined the practical uses of linear and nonlinear components for analyzing volatility measures in financial markets, particularly the swap option market. We see that the principal component extracts by

---

[1]For the three-year maturity the linear root mean squared error is slightly lower than the error of the nonlinear component. However, the slightly higher linear statistic is due to a few jumps in the nonlinear error. Otherwise, the nonlinear error remains much closer to zero. This explains the divergent results of the squared error and Diebold-Mariano statistics.

TABLE 9.5. U.S. Implied Volatility Estimates: Out-of-Sample Prediction Performance Root Mean Squared Error

	Maturity in Years					
	2	3	4	5	7	10
Linear	5.761	2.247	1.585	3.365	5.843	7.699
Nonlinear	1.575	2.249	2.423	2.103	1.504	1.207

	Diebold-Mariano Tests* Maturity in Years					
	2	3	4	5	7	10
DM-0	0.000	0.000	0.997	0.000	0.000	0.000
DM-1	0.000	0.002	0.986	0.000	0.000	0.000
DM-2	0.000	0.006	0.971	0.000	0.000	0.000
DM-3	0.000	0.011	0.956	0.000	0.000	0.000
DM-4	0.000	0.017	0.941	0.001	0.000	0.000

Note: *P-values
DM-0 to DM-4: tests at autocorrelations 0 to 4.

the nonlinear auto-associative mapping are much more effective for out-of-sample predictions than the linear component. However, both components, for both countries, follow broadly similar patterns. Doing a simple test of causality, we find that both the U.S. components, whether linear or nonlinear, can help predict the linear or nonlinear Hong Kong components, but not vice-versa. This should not be surprising, since the U.S. market is much larger and many of the pricing decisions would be expected to follow U.S. market developments.

9.3.1 MATLAB Program Notes

The main MATLAB program for this chapter is *neftci_capfloor_prog.m*. The final output and data are in *USHKCAPFLOOR_ALL_run77.mat*.

9.3.2 Suggested Exercises

An interesting extension would be to find one principal component for the combined set of U.S. and Hong Kong cap-floor volatilities. Following this, the reader could compare the one principal component for the combined set with the corresponding principal component for each country. Are there any differences?

Bibliography

Aarts, E., and J. Korst (1989), *Simulated Annealing and Boltzmann Machines: A Stochastic Approach to Combinatorial Optimization and Neural Computing*. New York: John Wiley and Sons.

Akaike, H. (1974), "A New Look At Statistical Model Identification," *IEEE Transactions on Automatic Control*, AC-19, 46: 716–723.

Altman, Edward (1981), *Applications of Classification Procedures in Business, Banking and Finance*. Greenwich, CT: JAI Press.

Arifovic, Jasmina (1996), "The Behavior of the Exchange Rate in the Genetic Algorithm and Experimental Economies," *Journal of Political Economy* 104: 510–541.

Bäck, T. (1996), *Evolutionary Algorithms in Theory and Practice*. Oxford: Oxford University Press.

Baesens, Bart, Rudy Setiono, Christophe Mues, and Jan Vanthienen (2003), "Using Neural Network Rule Extraction and Decision Tables for Credit-Risk Evaluation." *Management Science* 49: 312–329.

Banerjee, A, R.L. Lumsdaine, and J. H. Stock (1992), "Recursive and Sequential Tests of the Unit Root and Trend-Break Hypothesis: Theory and International Evidence," *Journal of Business and Economic Statistics* 10: 271–287.

Bates, David S. (1996), "Jumps and Stochastic Volatility: Exchange Rate Processes Implicit in Deutsche Mark Options," *Review of Financial Studies* 9: 69–107.

Beck, Margaret (1981), "The Effects of Seasonal Adjustment in Econometric Models." Discussion Paper 8101, Reserve Bank of Australia.

Bellman, R. (1961), *Adaptive Control Processes: A Guided Tour.* Princeton, NJ: Princeton University Press.

Beltratti, Andrea, Serio Margarita, and Pietro Terna (1996), *Neural Networks for Economic and Financial Modelling.* Boston: International Thomson Computer Press.

Beresteanu, Ariel (2003), "Nonparametric Estimation of Regression Functions under Restrictions on Partial Derivatives." Working Paper, Department of Economics, Duke University. Webpage: www.econ.duke.edu/~arie/shape.pdf.

Bernstein, Peter L. (1998), *Against the Gods: The Remarkable Story of Risk.* New York: John Wiley and Sons.

Black, Fisher, and Myron Sholes (1973), "The Pricing of Options and Corporate Liabilities," *Journal of Political Economy* 81: 637–654.

Bollerslev, Timothy (1986), "Generalized Autoregressive Conditional Heteroskedasticity," *Journal of Econometrics,* 31: 307–327.

——— (1987), "A Conditionally Heteroskedastic Time Series Model for Speculative Prices and Rates of Return," *Review of Economics and Statistics* 69: 542–547.

Breiman, Leo (1996), "Bagging Predictors," *Machine Learning* 24: 123–140.

Brock, W., W. Deckert, and J. Scheinkman (1987), "A Test for Independence Based on the Correlation Dimension," Working Paper, Department of Economics, University of Wisconsin at Madison.

———, and B. LeBaron (1996), "A Test for Independence Based on the Correlation Dimension." *Econometric Reviews* 15: 197–235.

Buiter, Willem, and Nikolaos Panigirtazoglou (1999), "Liquidity Traps: How to Avoid Them and How to Escape Them." Webpage: www.cepr.org/pubs/dps/DP2203.dsp.

Campbell, John Y., Andrew W. Lo, and A. Craig MacKinlay (1997), *The Econometrics of Financial Markets*. Princeton, NJ: Princeton University Press.

Carreira-Perpinan, M.A. (2001), *Continuous Latent Variable Models for Dimensionality Reduction*. University of Sheffield, UK: Ph.D. Thesis. Webpage: www.cs.toronto.edu/~miguel/papers.html.

Chen, Xiaohong, Jeffery Racine, and Norman R. Swanson (2001), "Semiparametric ARX Neural Network Models with an Application to Forecasting Inflation," *IEEE Transactions in Neural Networks* 12: 674–683.

Chow, Gregory (1960), "Statistical Demand Functions for Automobiles and Their Use for Forecasting," in Arnold Harberger (ed.), *The Demand for Durable Goods*. Chicago: University of Chicago Press, 149–178.

Clark, Todd E., and Michael W. McCracken (2001), "Tests of Forecast Accuracy and Encompassing for Nested Models," *Journal of Econometrics* 105: 85–110.

Clark, Todd E., and Kenneth D. West (2004), "Using Out-of-Sample Mean Squared Prediction Errors to Test the Martingale Difference Hypothesis." Madison, WI: Working Paper, Department of Economics, University of Wisconsin.

Clouse, James, Dale Henderson, Athanasios Orphanides, David Small, and Peter Tinsley (2003), "Monetary Policy when the Nominal Short Term Interest Rate is Zero," in *Topics in Macroeconomics*. Berkeley Electronic Press: www.bepress.com.

Collin-Dufresne, Pierre, Robert Goldstein, and J. Spencer Martin (2000), "The Determinants of Credit Spread Changes." Working Paper, Graduate School of Industrial Administration, Carnegie Mellon University.

Cook, Steven (2001), "Asymmetric Unit Root Tests in the Presence of Structural Breaks Under the Null," *Economics Bulletin*: 1–10.

Corradi, Valentina, and Norman R. Swanson (2002), "Some Recent Developments in Predictive Accuracy Testing with Nested and (Generic) Nonlinear Alternatives." New Brunswick, NJ: Working Paper, Department of Economics, Rutgers University.

Craine, Roger, Lars A. Lochester, and Knut Syrtveit (1999), "Estimation of a Stochastic-Volatility Jump Diffusion Model." Unpublished

Manuscript, Department of Economics, University of California, Berkeley.

Dayhoff, Judith E., and James M. DeLeo (2001), "Artificial Neural Networks: Opening the Black Box." *Cancer* 91: 1615–1635.

De Falco, Ivanoe (1998), "Nonlinear System Identification by Means of Evolutionarily Optimized Neural Networks," in Quagliarella, D., J. Periaux, C. Poloni, and G. Winter (eds.), *Genetic Algorithms and Evolution Strategy in Engineering and Computer Science: Recent Advances and Industrial Applications.* West Sussex, England: John Wiley and Sons.

Dickey, D.A., and W.A. Fuller (1979), "Distribution of the Estimators for Autoregressive Time Series With a Unit Root," *Journal of the American Statistical Association* 74: 427–431.

Diebold, Francis X., and Roberto Mariano (1995), "Comparing Predictive Accuracy," *Journal of Business and Economic Statistics*, 3: 253–263.

Engle, Robert (1982), "Autoregressive Conditional Heterskedasticity with Estimates of the Variance of United Kingdom Inflation," *Econometrica* 50: 987–1007.

————, and Victor Ng (1993), "Measuring the Impact of News on Volatility," *Journal of Finance* 48: 1749–1778.

Essenreiter, Robert (1996), *Geophysical Deconvolution and Inversion with Neural Networks.* Department of Geophysics, University of Karlsruhe, www-gpi.physik.uni-karlsruhe.de.

Evans, Martin D., and Paul D. McNelis (2000), "Student Evaluations and the Assessment of Teaching Effectiveness: What Can We Learn from the Data." Webpage: www.georgetown.edu/faculty/mcnelisp/Evans-McNelis.pdf.

Fotheringhame, David, and Roland Baddeley (1997), "Nonlinear Principal Components Analysis of Neuronal Spike Tran Data." Working Paper, Department of Physiology, University of Oxford.

Franses, Philip Hans, and Dick van Dijk (2000), *Non-linear Time Series Models in Empirical Finance.* Cambridge, UK: Cambridge University Press.

Gallant, A. Ronald, Peter E. Rossi, and George Tauchen (1992), "Stock Prices and Volume." *Review of Financial Studies* 5: 199–242.

Geman, S., and D. Geman (1984), "Stochastic Relaxation, Gibbs Distributions, and the Bayesian Restoration of Images," *IEEE Transactions on Pattern Analysis and Machine Intelligence*, PAMI-6: 721–741.

Genberg, Hans (2003), "Foreign Versus Domestic Factors as Sources of Macroeconomics Fluctuations in Hong Kong." HKIMR Working Paper 17/2003.

———, and Laurent Pauwels (2003), "Inflation in Hong Kong, SAR–In Search of a Transmission Mechanism." HKIMR Working Paper No. 01/2003.

Gerlach, Stefan, and Wensheng Peng (2003), "Bank Lending and Property Prices in Hong Kong." Hong Kong Institute of Economic Research, Working Paper 12/2003.

Giordani, Paolo (2001) "An Alternative Explanation of the Price Puzzle." Stockholm: Sveriges Riksbank Working Paper Series No. 125.

Goodhard, Charles, and Boris Hofmann (2003), "Deflation, Credit and Asset Prices." HKIMR Working Paper 13/2003.

Goodfriend, Marvin (2000), "Overcoming the Zero Bound on Interest Rate Policy," *Journal of Money, Credit, and Banking* 32: 1007–1035.

Greene, William H. (2000), *Econometric Analysis*. Upper Saddle River, NJ: Prentice Hall.

Granger, Clive W.J., and Yongil Jeon (2002), "Thick Modeling." Unpublished Manuscript, Department of Economics, University of California, San Diego, *Economic Modeling*, forthcoming.

Ha, Jimmy, and Kelvin Fan (2002), "Price Convergence Between Hong Kong and the Mainland." Hong Kong Monetary Authority Research Memoranda.

Hannan, E.J., and B.G. Quinn (1979), "The Determination of the Order of an Autoregression," *Journal of the Royal Statistical Society* B, 41: 190–195.

Hansen, Lars Peter, and Thomas J. Sargent (2000), "Wanting Robustness in Macroeconomics." Manuscript, Department of Economics, Stanford University. Website: www.stanford.edu/~sargent.

Hamilton, James D. (1989), "A New Approach to the Economic Analysis of Nonstationary Time Series Subject to Changes in Regime," *Econometrica* 57: 357–384.

——— (1990), "Analysis of Time Series Subject to Changes in Regime," *Journal of Econometrics* 45: 39–70.

——— (1994), *Times Series Analysis*. Princeton, NJ: Princeton University Press.

Harvey, D, S. Leybourne, and P. Newbold (1997), "Testing the Equality of Prediction Mean Squared Errors," *International Journal of Forecasting* 13: 281–291.

Haykin, Simon (1994) *Neural Networks: A Comprehensive Foundation.* Saddle River, NJ: Prentice-Hall.

Heer, Burkhard, and Alfred Maussner (2004), *Dynamic General Equilibrium Modelling-Computational Methods and Applications.* Berlin: Springer Verlag. Forthcoming.

Hess, Allan C. (1977), "A Comparison of Automobile Demand Functions," *Econometrica* 45: 683–701.

Hoffman, Boris (2003), "Bank Lending and Property Prices: Some International Evidence." HKIMR Working Paper 22/2003.

Hornik, K., X. Stinchcomb, and X. White (1989), "Multilayer Feedforward Networks are Universal Approximators." *Neural Net* 2: 359–366.

Hsieh, D., and B. LeBaron (1988a), "Small Sample Properties of the BDS Statistic, I," in W. A. Brock, D. Hsieh, and B. LeBaron (eds.), *Nonlinear Dynamics, Chaos, and Stability.* Cambridge, MA: MIT Press.

——— (1988b), "Small Sample Properties of the BDS Statistic, II," in W. A. Brock, D. Hsieh, and B. LeBaron (eds.), *Nonlinear Dynamics, Chaos, and Stability.* Cambridge, MA: MIT Press.

——— (1988c), "Small Sample Properties of the BDS Statistic, III," in W. A. Brock, D. Hsieh, and B. LeBaron (eds.), *Nonlinear Dynamics, Chaos, and Stability.* Cambridge, MA: MIT Press.

Hutchinson, James M., Andrew W. Lo, and Tomaso Poggio (1994), "A Nonparametric Approach to Pricing and Hedging Derivative Securities Via Learning Networks," *Journal of Finance* 49: 851–889.

Ingber, L. (1989), "Very Fast Simulated Re-Annealing," *Mathematical Computer Modelling* 12: 967–973.

Issing, Othmar (2002), "Central Bank Perspectives on Stabilization Policy." *Federal Reserve Bank of Kansas City Economic Review*, 87: 15–36.

Jarque, C.M., and A.K. Bera (1980), "Efficient Tests for Normality, Homoskedasticity, and Serial Independence of Regression Residuals," *Economics Letters* 6: 255–259.

Judd, Kenneth L. (1998), *Numerical Methods in Economics*. Cambridge, MA: MIT Press.

Kantz, H., and T. Schreiber (1997), *Nonlinear Time Series Analysis*. Cambridge, UK: Cambridge University Press.

Kirkpatrick, S, C.D. Gelatt Jr., and M.P. Vecchi (1983), "Optimization By Simulated Annealing," *Science* 220: 671–680.

Kočenda, E. (2001) An Alternative to the BDS Test: Integration Across the Correlation Integral. Econometric Reviews 20, 337–351.

Krugman, Paul (1998), "Special Page on Japan: Introduction." Webpage: web.mit.edu/krugman/www/jpage.html.

Kuan, Chung-Ming, and Halbert White (1994), "Artifical Neural Networks: An Econometric Perspective," *Econometric Reviews* 13: 1–91.

Kuan, Chung-Ming, and Tung Liu (1995), "Forecasting Exchange Rates Using Feedforward and Recurrent Neural Networks," *Journal of Applied Econometrics* 10: 347–364.

Lai, Tze Leung, and Samuel Po-Shing Wong (2001), "Stochastic Neural Networks with Applications to Nonlinear Time Series." *Journal of the American Statistical Association* 96: 968–981.

LeBaron, Blake (1998), "An Evolutionary Bootstrap Method for Selecting Dynamic Trading Stratergies", in A.-P. N. Refenes, A.N. Burgess and J.D. Moody (eds.), *Decision Technologies for Computational Finance*, Ansterdam: Kluwer Academic Publishers, 141–160.

Lee, T.H., H. White, and C.W.J. Granger (1992), "Testing for Neglected Nonlinearity in Times Series Models: A Comparison of Neural Network Models and Standard Tests," *Journal of Econometrics* 56: 269–290.

Ljung, G.M., and G.E.P. Box (1978), "On a Measure of Lack of Fit in Time Series Models." *Biometrika* 65: 257–303.

Lumsdaine, Robin L., and D. H. Papell (1997), "Multiple Trend Breaks and the Unit Root Hypothesis," *Review of Economics and Statistics*: 212–218.

Mandic, Danilo, and Jonathan Chambers (2001), *Recurrent Neural Networks for Prediction: Learning Algorithms, Architectures, and Stability*. New York: John Wiley and Sons.

McCarthy, Patrick S. (1996), "Market Price and Income Elasticities of New Vehicles," *Review of Economics and Statistics* 78: 543–548.

McKibbin, Warwick (2002), "Macroeconomic Policy in Japan," *Asian Economic Paper* 1: 133–169.

———, and Peter Wilcoxen (1998), "The Theoretical and Empirical Structure of the G-Cubed Model," *Economic Modelling* 16: 123–148.

McLeod, A. I., and W.K. Li (1983), "Diagnostic Checking of ARMA Time Series Models Using Squared-Residual Autocorrelations," *Journal of Time Series Analysis* 4: 269–273.

McNelis, P., and G. Nickelsburg (2002), "Forecasting Automobile Production in the United States." Manuscript, Economics Dept., Georgetown University.

McNelis, Paul D., and Peter McAdam (2004), "Forecasting Inflation with Thick Models and Neural Networks." Working Paper 352, European Central Bank. Webpage: www.ecb.int/pub/wp/ecbsp352.pdf.

Meltzer, Alan (2001), "Monetary Transmission at Low Inflation: Some Clues from Japan," *Monetary and Economic Studies* 19(S-1): 13–34.

Merton, Robert (1973), "An Intertemporal Capital Asset Pricing Model." *Econometrica* 41: 867–887.

Metropolis, N., A.W. Rosenbluth, M. N. Rosenbluth, A.H. Teller, and E. Teller (1953), "Equation of State Calculations by Fast Computing Machines," *Journal of Chemical Physics* 21: 1087–1092.

Michalewicz, Zbigniew (1996), *Genetic Algorithms + Data Structures = Evolution Programs*. Third Edition. New York: Springer-Verlag.

———, and David B. Fogel (2002), *How to Solve It: Modern Heuristics*. New York: Springer-Verlag.

Miller, W. Thomas III, Richard S. Sutton, and Paul J. Werbos (1990), *Neural Networks for Control*. Cambridge, MA: MIT Press.

Neftçi, Salih (2000), *An Introduction to the Mathematics of Financial Derivatives*. San Diego, CA: Academic Press.

Perron, Pierre (1989), "The Great Crash, the Oil Price Shock, and the Unit Root Hypothesis," *Econometrics* 57: 1361–1401.

Pesaran, M.H., and A. Timmermann (1992), "A Simple Nonparametric Test of Predictive Performance," *Journal of Business and Economic Statistics* 10: 461–465.

Qi, Min (1999), "Nonlinear Predictability of Stock Returns Using Financial and Economic Variables," *Journal of Business and Economics Statistics* 17: 419–429.

Quagliarella, Domenico, and Alessandro Vicini (1998), "Coupling Genetic Algorithms and Gradient Based Optimization Techniques," in Quagliarella, D., J. Periaux, C. Poloni, and G. Winter (eds.), *Genetic Algorithms and Evolution Strategy in Engineering and Computer Science: Recent Advances and Industrial Applications.* West Sussex, England: John Wiley and Sons, Ltd.

Quagliarella, D., J. Periaux, C. Poloni, and G. Winter (1998), *Genetic Algorithms and Evolution Strategy in Engineering and Computer Science: Recent Advances and Industrial Applications.* West Sussex, England: John Wiley and Sons, Ltd.

Razzak, Weshah A. "Wage-Price Dynamics, the Labor Market, and Deflation in Hong Kong." HKIMR Working Paper 24/2003.

Rissanen, J. (1986a), "A Predictive Least-Squares Principle," *IMA Journal of Mathematical Control and Information* 3: 211–222.

———— (1986b), "Stochastic Complexity and Modeling," *Annals of Statistics* 14: 1080–1100.

Robinson,   Guy   (1995),   "Simulated   Annealing."   Webpage: www.npac.syr.edu/ copywrite/pcw/node252.

Ross, S. (1976), "The Arbitrage Theory of Capital Asset Pricing," *Journal of Economic Theory* 13: 341–360.

Rustichini, Aldo, John Dickhaut, Paolo Ghirardato, Kip Smith, and Jose V. Pardo (2002), "A Brain Imaging Study of Procedural Choice," Working Paper, Department of Economics, University of Minnesota. Webpage: http://www.econ.umn.edu/~arust/ProcCh3.pdf.

Sargent, Thomas J. (1997), *Bounded Rationalilty in Macroeconomics.* Oxford: Oxford University Press.

———— (1999), *The Conquest of American Inflation.* Princeton, NJ: Princeton University Press.

Schwarz, G. (1978), "Estimating the Dimension of a Model," *Annals of Statistics* 6: 461–464.

Sims, Christopher (1992), "Interpreting the Macroeconomic Times Series Facts: The Effects of Monetary Policy." *European Economic Review* 36: 2–16.

———, and Mark W. Watson (1998), "A Comparison of Linear and Nonlinear Univariate Models for Forecasting Macroeconomic Time Series." Cambridge, MA: National Bureau of Economic Research Working Paper 6607. Website: www.nber.org/papers/w6607.

Stock, James H., and Mark W. Watson (1999), "Forecasting Inflation," *Journal of Monetary Economics* 44: 293–335.

Sundermann, Erik (1996), "Simulated Annealing." Webpage: petaxp.rug. ac.be/~erik/research/research-part2.

Svensson, Lars E. O., (2003), "Escaping from a Liquidity Trap and Deflation: The Foolproof Way and Others," *Journal of Economic Perspectives*.

Teräsvirta, T. (1994), "Specification, Estimation, and Evaluation of Smooth-Transition Autogressive Models," *Journal of the American Statistical Association* 89: 208–218.

———, and H.M. Anderson (1992), "Characterizing Nonlinearities in Business Cycles Using Smooth Transition Autoregressive Models," *Journal of Applied Econometrics* 7: S119–S136.

van Dijk, Dick, Timo Teräsvirta, and Philip Hans Franses (2000), "Smooth Transition Autoregressive Models—A Survey of Recent Developments." Research Report EI2000–23A. Rotterdam: Erasmus University, Econometric Institute.

Tsay, Ruey S. (2002), *Analysis of Financial Time Series*. New York: John Wiley and Sons, Inc.

van Laarhoven, P.J.M., and E.H.L. Aarts (1988), *Simulated Annealing: Theory and Applications*. Boston, MA: Kluwer Academic Publishers.

Werbos, Paul John (1994), *The Roots of Backpropagation: From Ordered Derivatives to Neural Networks and Political Forecasting*. New York: Wiley Interscience.

White, Halbert (1980), "A Heteroskedasticity Covariance Matrix and a Direct Test for Heteroskedasticity." *Econometrica* 48: 817–838.

Wolkenhauer, Olaf (2001), *Data Engineering.* New York: John Wiley and Sons.

Yoshino, Naoyuki and Eisuke Sakakibara (2002), "The Current State of the Japanese Economy and Remedies," *Asian Economic Papers* 1: 110–126.

Zivot, E., and D.W.K. Andrews (1992), "Further Evidence on the Great Crash, the Oil Price Shock, and the Unit-Root Hypothesis," *Journal of Business and Statistics* 10: 251–270.

Index

233